iPhone™ 3G
PORTABLE GENIUS

iPhone™ 3G

PORTABLE GENIUS

by Paul McFedries and David Pabian

WILEY

Wiley Publishing, Inc.

iPhone™ 3G Portable Genius

Published by
Wiley Publishing, Inc.
10475 Crosspoint Blvd.
Indianapolis, IN 46256
www.wiley.com

Copyright © 2008 by Wiley Publishing, Inc., Indianapolis, Indiana

Published simultaneously in Canada

ISBN: 978-0-470-42348-6

Manufactured in the United States of America

10 9 8 7 6 5 4 3 2 1

For general information on our other products and services or to obtain technical support, please contact our Customer Care Department within the U.S. at (800) 762-2974, outside the U.S. at (317) 572-3993 or fax (317) 572-4002.

Wiley also publishes its books in a variety of electronic formats. Some content that appears in print may not be available in electronic books.

Library of Congress Control Number is available from the publisher.

WILEY

About the Authors

Paul McFedries is a Mac expert and full-time technical writer. Paul has been authoring computer books since 1991 and has more than 60 books to his credit. Paul's books have sold more than three million copies worldwide. These books include the Wiley titles *Teach Yourself VISUALLY Macs*, *Teach Yourself VISUALLY Computers, Fifth Edition*, and *The Unofficial Guide to Microsoft Office 2007*. Paul is also the proprietor of Word Spy (www.wordspy.com), a Web site that tracks new words and phrases as they enter the language.

David Pabian is a freelance writer and Mac enthusiast. A self-taught Mac expert and early adopter, he can often be found troubleshooting all things Mac for his family and friends.

Credits

Senior Acquisitions Editor
Stephanie McComb

Senior Project Editor
Cricket Krengel

Technical Editor
G. Smith

Copy Editor
Kim Heusel

Editorial Manager
Robyn B. Siesky

Vice President & Group Executive Publisher
Richard Swadley

Vice President & Publisher
Barry Pruett

Business Manager
Amy Knies

Senior Marketing Manager
Sandy Smith

Special Help
Jarod Brock

Composition Management
Clint Lahnen
Shelley Lea
Kathie Rickard
Debbie Stailey

Project Coordinator
Kristie Rees

Graphics and Production Specialists
Elizabeth Brooks
Laura Campbell
Ronda David-Burroughs
Jennifer Henry
Andrea Hornberger
Brent Savage

Quality Control Technicians
Laura Albert
David Faust
Caitie Kelly
Rob Springer
Amanda Steiner

Proofreading
Shannon Ramsey

Indexing
Johnna VanHoose

Acknowledgments

If you think *using* the iPhone 3G is a blast, you should try about it! We had so much fun poking and prodding the iPhone 3G to see just what it's capable of, that we're worried our publisher won't pay us next time! Adding to the fun were the great people at Wiley that we got to work with. They included Acquisition Editor Stephanie McComb, who was kind enough (or was it brave enough?) to ask us to write the book; Project Editor Cricket Krengel, whose unflappable calm under intense pressure is beyond admirable; and Copy Editor Kim Heusel, whose hawk-like attention to the all-important details made this a much better book. Many heartfelt thanks to all of you for outstanding work on this project.

~ Paul McFedries

First, I want to thank my entire family. I truly have the greatest family and I owe them more than just my gratitude. To my parents, Janelle and Mike, thanks for all of the support over the years and teaching me just about everything worth knowing. I appreciate it more than you will ever know. Thanks to my brother and sister, Dan and Sarah, for getting on my case when I deserve it, and for always being willing to hang out. Tucker and Dewey, thanks for spending some of the late nights with me even if you were only able to offer up blank stares as inspiration. Mandy, thanks for all the food and conversation. It would have been much less pleasant without you, to say the least. Thanks Taylor and Cameron, for making me laugh and keeping me company.

Shilpa, thanks for reminding me that there is a world out there. I'm not sure I would have left the house if not for you. Thanks for keeping me company while writing this. To Scott, keep fighting the good fight. Thanks Erik, for teaching me to think outside the box. Daniel, thanks for always being willing to sit on the porch just a little bit longer.

Finally, thanks to everyone who worked on this book. The difference between what left my hands and what ended up on the shelf is amazing.

I'm extremely grateful to Paul McFedries for all his fast and exceptional work, and to Jarod Brock for weathering the storm, all night under an awning just to make sure we got a phone.

~ David Pabian

Contents

chapter 3

How Can I Make the Most of Web
Surfing with My iPhone 3G? 50

chapter 4

How Do I Maximize E-mail on
My iPhone 3G? 74

chapter 5

chapter 6

Introduction

We define *technolust* as an overwhelming and unstoppable desire to own some new gadget or high-tech toy. The iPhone 3G with its built-in GPS receiver, App Store, flush stereo jack, improved battery life, and, at long last, fast 3G network support, takes technolust to a new level. That's probably why *you* have one, and chances are you're as thrilled with it as we are.

However, lust is a short-term state, and inevitably those feelings fade as you settle into the daily routine of calling, surfing, e-mailing, texting, and scheduling. That's when you take a step back and say, "Whoa, dude, this is one fancy phone; it's gotta be able to do more than that!" It's also about this time when you begin to notice the iPhone 3G's flaws. The Home screen icons aren't arranged in a way that makes sense for you; getting a Bluetooth headset to connect is a bit of a mystery; you've downloaded some applications and sometimes they just freeze up on you.

Yes, the iPhone 3G is mighty simple to use out-of-the-box, but some of its most useful and powerful features are hidden away in obscure parts of the system. Sure, the iPhone 3G doesn't get in your way when you're trying to be productive or creative, but sometimes it does something (or forces you to do something) that just makes you want to scratch your head in wonderment. Sure, the iPhone 3G's robust design makes it a reliable device day after day, but even the best-built machines can have problems.

When you come upon the iPhone 3G's dark side, you might consider making an appointment with your local Apple Store's Genius Bar, and more often than not the on-duty genius gives you good advice on how to overcome the iPhone 3G's limitations, work around its annoyances, and fix its failures. The Genius Bar is a great thing, but it isn't exactly a convenient thing. You can't just drop by to get help; you have to make an appointment; you have to drag yourself down to the store, perhaps wait for your genius, get the problem looked at, and then make your way back home; and in some cases you may need to leave your iPhone 3G for a while (No!) to get the problem checked out and hopefully resolved.

What you really need is a version of the Genius Bar that's easier to access, more convenient, and doesn't require tons of time or leaving your iPhone 3G in the hands of a stranger. What you really need is a "portable" genius that enables you to be more productive and solve problems wherever you and your iPhone 3G happen to be.

Welcome, therefore, to *iPhone 3G Portable Genius*. This book is like a mini Genius Bar all wrapped up in an easy-to-use, easy-to-access, and eminently portable format. In this book, you learn how to get more out of your iPhone 3G by learning how to access all the really powerful and timesaving features that aren't obvious at a casual glance. In this book, you learn how to avoid your iPhone 3G's more annoying character traits and, in those cases where such behavior can't be avoided, you learn how to work around it. In this book, you learn how to prevent iPhone 3G problems from occurring, and just in case your preventative measures are for naught, you learn how to fix many common problems yourself.

This book is for iPhone 3G users who know the basics but want to take their iPhone 3G education to a higher level. It's a book for people who want to be more productive, more efficient, more creative, and more self-sufficient (at least as far as the iPhone 3G goes). It's a book for people who use their iPhone 3G every day, but would like to incorporate it into more of their day-to-day activities. It's a book we had a blast writing, so we think it's a book you'll enjoy reading.

The iPhone 3G is justly famous for its stylish, curvaceous design and for its slick, effortless touchscreen. However, although good looks and ease of use are important for any smartphone, it's what you *do* with that phone that's important. The iPhone 3G helps by offering lots of features, but chances are those features aren't set up to suit the way you work. Maybe your most-used Home screen icons aren't at the top of the screen where they should be, or perhaps your iPhone 3G goes to sleep too soon. This chapter shows you how to configure your iPhone 3G to solve these and many other annoyances so the phone works the way you do.

Customizing the Home Screen to Suit Your Style

The Home screen is your starting point for all things iPhone 3G, and what could be simpler? Just tap the icon you want and the application loads lickety-split. Ah, but things are never so simple, are they? In fact, there are a couple of hairs in the Home screen soup:

- The icons in the top row are a bit easier to find and a bit easier to tap.

- If you have more than 16 icons, they extend onto a second (or third or fourth) Home screen. If the application you want isn't on the main Home screen, you must first flick to the screen that has the application's icon (or tap its dot) and then tap the icon.

Note

How do you end up with more than 16 icons? Easy: the App Store. This is an online retailer solely devoted to applications designed to work with the iPhone 3G's technologies: multi-touch, GPS, the accelerometer, wireless, and more. You can download applications via your cellular network or your Wi-Fi connection, so you can always get applications when you need them. In the Home screen, tap the App Store icon to see what's available.

All this means that you can make the Home screen more efficient by moving your four most-used icons to the top row of the main Home screen, and make sure that any icon you tap frequently appears somewhere on the main Home screen. You can do all this by rearranging the Home screen icons as follows:

1. **Display the Home screen.**

2. **Tap and hold any Home screen icon.** When you see the icons wiggling, release your finger.

3. **Tap and drag the icons into the positions you prefer.**

4. **Press the Home button.** iPhone 3G saves the new icon arrangement.

Note

The icons in the Home screen's menu bar are also fair game, and you can drag them left and right to change the order. Unfortunately, you can't replace the menu bar icons with any other Home screen icons.

Moving unused icons off the main Home screen

The best way to make the main Home screen more manageable is to get rid of any icons you don't use. Not an investor? Get rid of the Stocks icon! No use for arithmetic? Say so long to the Calculator icon! Installed a bunch of applications you use only rarely? Get rid of them too!

Unfortunately, you can't delete the default iPhone 3G icons, and although you can uninstall any third-party applications, you probably don't want to go that far for any application you still use once in a while. The solution to both problems is to create a second Home screen and move your little-used icons to that screen. That way, your main Home screen holds just your favorite icons, and the ones you use once in a blue moon (or never) are out of the way.

Here are the steps to follow:

1. **In the Home screen, tap and hold any Home screen icon until you see all the icons wiggling**.

2. **For each icon you want off the main Home screen, tap and drag the icon to the right until the new Home screen appears, and then release the icon.**

3. **Press the Home button.** iPhone 3G saves the new icon arrangement.

Adding a Safari Web Clip to the Home screen

Do you have a Web page that you visit all the time? You can set up that page as a bookmark in iPhone 3G's Safari browser, but there's an even faster way to access the page: add it to the Home screen as a Web Clip icon. A *Web Clip* is a link to a page that preserves the page's scroll position and zoom level. For example, suppose a page has a form at the bottom. To use that form, you have to navigate to the page, scroll to the bottom, and then zoom in to the form to see it better. However, you can perform all three actions — navigate, scroll, and zoom — automatically with a Web Clip.

Follow these steps to save a page as a Web Clip icon on the Home screen:

1. **Use your iPhone 3G's Safari browser to navigate to the page you want to save.**

2. **Scroll to the portion of the page you want to see.**

3. **Pinch and spread your fingers over the area you want to zoom in on until you can comfortably read the text.**

4. **Press + at the bottom of the screen.** iPhone 3G displays a list of options.

5. **Tap Add to Home Screen.** iPhone 3G prompts you to edit the Web Clip name, as shown in figure 1.1.

6. **Edit the name as needed.** Names up to about 10-14 characters can display on the Home screen without being broken. (The fewer uppercase letters you use, the longer the name can be.) For longer names, iPhone 3G displays the first few and last few characters (depending on the locations of spaces in the name), separated by an ellipsis (...). For example, if the name is My Home Page, it appears in the Home screen as My Ho...Page

7. **Tap Add.** iPhone 3G adds the Web Clip to the Home screen and displays the Home screen. Figure 1.2 shows the Home screen with a Web Clip added.

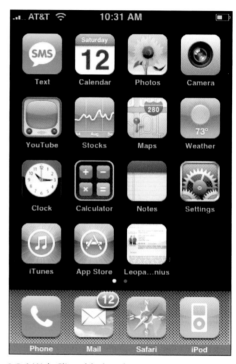

1.1 You can edit the Web Clip name before adding the icon to the Home screen.

1.2 A Web Clip added to the Home screen.

Genius

To delete a Web Clip from the Home screen, tap and hold any Home screen icon until the icon dance begins. Each Web Clip icon displays an X in the upper-left corner. Tap the X of the Web Clip you want to remove. When iPhone 3G asks you to confirm, tap Delete, and then press the Home button to save the configuration.

Resetting the default Home screen layout

If you make a bit of a mess of your Home screen, or if someone else is going to be using your iPhone 3G, you can reset the Home screen icons to their default layout. Follow these steps:

1. **On the Home screen, tap Settings.** The Settings application appears.

2. **Tap General.** The General screen appears.

3. **Scroll down and tap Reset.** The Reset screen appears.

4. **Tap Reset Home Screen Layout.** iPhone 3G warns you that the Home screen will be reset to the factory default layout.

5. **Tap Reset Home Screen.** iPhone 3G resets the home screen to the default layout, but it doesn't delete added Application buttons.

Protecting iPhone 3G with a Passcode

When your iPhone 3G is asleep, the phone is locked in the sense that tapping the touchscreen or pressing the volume controls does nothing. This sensible arrangement prevents accidental taps when the phone is in your pocket or rattling around in your backpack or handbag. To unlock the phone, you either press the Home button or the Sleep/Wake button, drag the Slide to Unlock slider, and you're back in business.

Unfortunately, this simple technique means that anyone else who gets his or her mitts on your iPhone 3G can also be quickly back in business — *your* business! If you have sensitive or confidential information on your phone, or if you want to avoid digital joyrides that run up massive roaming or data charges, you need to truly lock your iPhone 3G.

You do that by specifying a four-digit passcode that must be entered before anyone can use the iPhone 3G. Follow these steps to set up your passcode:

1. **On the Home screen, tap Settings.** The Settings application appears.

2. **Tap General.** The General screen appears.

Caution

You really, really need to remember your iPhone 3G passcode. If you forget it, you are locked out of your own phone and the only way to get back in is to completely reset the iPhone 3G (as described later in this chapter).

3. **Tap Passcode Lock.** The Set Passcode screen appears, as shown in figure 1.3.

4. **Tap your four-digit passcode.** For security, the numbers appear in the Enter a passcode box as dots. When you finish, iPhone 3G prompts you to reenter the passcode.

5. **Tap your four-digit passcode again.**

With your passcode now active, iPhone 3G displays the Passcode Lock screen. (You can also get to this screen by tapping Settings in the Home screen, then General, then Passcode Lock.) This screen offers four buttons:

- **Turn Passcode Off.** If you want to stop using your passcode, tap this button, and then enter the passcode (for security; otherwise an interloper could just shut off the passcode).

- **Change Passcode.** Tap this button to enter a new passcode. (Note that you first need to enter your old passcode, and then enter the new passcode.)

1.3 Use the Set Passcode screen to lock your iPhone 3G with a four-digit passcode.

- **Require Passcode.** This setting determines how much time elapses before the iPhone 3G locks the phone and requests the passcode. The default setting is Immediately, which means you see the Enter Passcode screen (see figure 1.4) as soon as you finish dragging Slide to Unlock. The other options are After 1 minute, After 5 minutes, After 15 minutes,

Note

If an emergency arises and you need to make a call for help, you probably don't want to mess around entering a passcode. Similarly, if something happens to you, another person who doesn't know your passcode may need to use your iPhone 3G to call for assistance. In both cases, you can temporarily bypass the passcode by tapping the Emergency Call button on the Enter Passcode screen.

After 1 hour, and After 4 hours. Use one of these settings if you want to be able to work with your iPhone 3G for a bit before getting locked out. For example, the After 1 minute option is good if you need to quickly check e-mail without having to enter your passcode.

- ⦿ **Show SMS Preview.** When this setting in On, iPhone 3G still displays a preview of an incoming SMS text message when it's locked; if you prefer not to show SMS text message previews when iPhone 3G is locked, change this setting to Off.

With the passcode activated, when you bring the iPhone 3G out of standby, you drag the Slide to Unlock slider as usual, and then the Enter Passcode screen appears, as shown in figure 1.4. Type your passcode to unlock the iPhone 3G.

1.4 To unlock your iPhone 3G, you need to enter your four-digit passcode.

Configuring When iPhone 3G Goes to Sleep

You can put your iPhone 3G into Standby mode at any time by pressing the Sleep/Wake button once. This drops the power consumption considerably (mostly because it shuts off the screen), but you can still receive incoming calls and text messages, and if you have the iPhone 3G's iPod application running, it continues to play.

However, if your iPhone 3G is on but you're not using it, the phone automatically goes into standby mode after two minutes. This is called Auto-Lock and it's a handy feature because it saves battery power (and prevents accidental taps) when your iPhone 3G is just sitting there.

If you're not comfortable with the default 2-minute Auto-Lock interval, you can make it shorter or longer, or you can disable it altogether. Here are the steps to follow:

1. **On the Home screen, tap Settings.** The Settings application appears.

2. **Tap General.** The General screen appears.

3. **Tap Auto-Lock.** The Auto-Lock screen appears, as shown in figure 1.5.

4. **Tap the interval you want to use.** To leave Auto-Lock activated, tap one of the following intervals: 1 minute, 2 Minutes, 3 Minutes, 4 Minutes, or 5 Minutes. To disable Auto-Lock, tap Never.

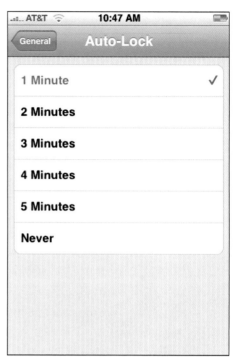

.ıl. AT&T 10:47 AM

General **Auto-Lock**

1 Minute ✓

2 Minutes

3 Minutes

4 Minutes

5 Minutes

Never

1.5 Use the Auto-Lock screen to set the Auto-Lock interval or to turn it off.

Turning Sounds On and Off

Your iPhone 3G is often a noisy little thing that makes all manner of rings, beeps, and boops, seemingly at the slightest provocation. Consider a short list of the events that can give the iPhone 3G's lungs a workout:

- Incoming calls
- Incoming e-mail messages
- Outgoing e-mail messages
- Incoming text messages
- New voicemail messages
- Calendar alerts
- Locking and unlocking the phone
- Tapping the keys on the onscreen keyboard

What a racket! None of this may bother you when you're on your own, but if you're in a meeting, a movie, or anywhere else where extraneous sounds are unwelcome, you might want to turn off some or all of the iPhone 3G's sound effects.

First, you should know that when a call comes in and you press the Sleep/Wake button once, your iPhone 3G silences the ringer. That's a sweet and useful feature, but the problem is that it may take you one or two rings before you can dig out your iPhone 3G and press Sleep/Wake, and by that time the folks nearby are already glaring at you.

To prevent this phone faux pas, you can switch your iPhone 3G into silent mode, which means it doesn't ring, and it doesn't play any alerts or sound effects. When the sound is turned off, only alarms that you've set using the Clock application sound. The phone still vibrates unless you turn this feature off as well. You switch the iPhone 3G between ring and silent modes using the Ring/Silent switch, which is located on the left side panel of the iPhone 3G, near the top. Use the following techniques to switch between silent and ring modes:

- To put the phone in silent mode, flick the Ring/Silent switch toward the back of the phone. You see a little orange dot on the switch and the iPhone 3G screen displays a bell with a slash through it. Your iPhone 3G is now in silent mode.

- To resume the normal ring mode, flick the Ring/Silent switch toward the front of the phone. The iPhone 3G screen displays a bell. Your iPhone 3G is now in normal ring mode.

If silent mode is a bit too drastic, you can control exactly which sounds your iPhone 3G utters by following these steps:

1. **On the Home screen, tap Settings.** The Settings application appears.

2. **Tap Sounds.** The Sounds screen appears, as shown in figure 1.6.

3. **In the Silent section, the Vibrate setting determines whether iPhone 3G vibrates when the phone is in silent mode.** Vibrating is a good idea in silent mode, so On is a good choice here.

1.6 Use the Sounds screen to turn the iPhone 3G's sounds on and off.

4. **In the Ring section, the Vibrate setting determines whether iPhone 3G vibrates when the phone is in ring mode.** Vibrating probably isn't all that important in ring mode, so feel free to change this setting to Off. The exception is if you reduce the ringer volume (see step 5), in which case setting Vibrate to On might help you notice an incoming call.

5. **Drag the volume slider to set the volume of the ringtone that plays when a call comes in.**

6. **To set a different default ringtone, tap Ringtone to open the Ringtone screen, tap the ringtone you want to use (iPhone 3G plays a preview), and then tap Sounds to return to the Sounds screen.**

7. **To set a different incoming text message sound, tap New Text Message to open the New Text Message screen, tap the sound effect you want to use (iPhone 3G plays a preview), and then tap Sounds to return to the Sounds screen.**

Note   If you don't want your iPhone 3G to play a sound when a new text message arrives, tap None in the New Text Message screen.

8. **For each of the following settings, tap the On/Off button to turn the sounds on or off:**

   ○ New Voicemail

   ○ New Mail

   ○ Sent Mail

   ○ Calendar Alerts

   ○ Lock Sounds

   ○ Keyboard Clicks

Adjusting the Brightness of the Screen

Your iPhone 3G's touchscreen offers a crisp, bright display that's easy to read in most situations. Unfortunately, keeping the screen bright enough to read comfortably extracts a heavy cost in battery power. To help balance screen brightness and battery life, your iPhone 3G comes with a built-in ambient light sensor. That sensor checks the surrounding light levels and adjusts the brightness of the iPhone 3G screen accordingly:

● If the ambient light is dim, the iPhone 3G screen is easier to read, so the sensor dims the screen brightness to save battery power.

● If the ambient light is bright, the iPhone 3G screen is harder to see, so the sensor brightens the screen to improve readability.

This feature is called Auto-Brightness, and it's sensible to let your iPhone 3G handle this stuff for you. However, if you're not happy with how Auto-Brightness works, or if you simply have an uncontrollable urge to tweak things, you can follow these steps to adjust the screen brightness by hand:

1. **On the Home screen, tap Settings.** The Settings application appears.

2. **Tap Brightness.** The Brightness screen appears, as shown in figure 1.7.

3. **Drag the Brightness slider left (for a dimmer screen) or right (for a brighter screen).**

4. **To prevent iPhone 3G from controlling the brightness automatically, turn the Auto-Brightness setting to Off.**

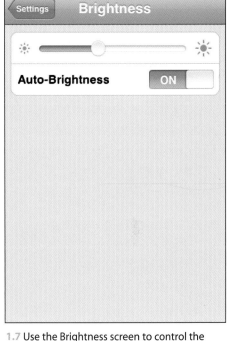

1.7 Use the Brightness screen to control the iPhone 3G's screen brightness by hand.

Note

Even if you leave Auto-Brightness turned on, you still might want to adjust the Brightness slider because this affects the relative brightness of the screen. For example, suppose you adjust the slider to increase brightness by 50 percent and you leave Auto-Brightness turned on. In this case, Auto-Brightness still adjusts the screen automatically, but any brightness level it chooses is 50 percent brighter than it would be otherwise.

Setting the iPhone 3G Wallpaper

The iPhone 3G wallpaper is the background image you see when you unlock the phone. That is, it's the image you see when the Slide to Unlock screen appears, and also when the Enter Passcode screen appears, if you're protecting your iPhone 3G with a passcode (as described earlier in this chapter). The default wallpaper is a photo of the Earth taken from space, and as nice as that photo is, you might just be getting a bit tired of looking at it. No worries! Your iPhone 3G comes with 15 other wallpapers you can choose, and you can even use one of your own photos as the wallpaper.

Using a predefined wallpaper

Here are the steps to follow to use one of iPhone 3G's predefined wallpapers:

1. **On the Home screen, tap Settings.** The Settings application appears.

2. **Tap Wallpaper.** The Wallpaper screen appears.

3. **Tap Wallpaper.** iPhone 3G displays its collection of wallpaper images, as shown in figure 1.8.

4. **Tap the image you want to use.** The Wallpaper preview screen appears.

5. **Tap Set Wallpaper.** iPhone 3G sets the image as the wallpaper.

Using an existing photo as the wallpaper

If you have images in your iPhone 3G's Camera Roll, or in a photo album synced from your computer, you can use one of those images as your wallpaper by following these steps:

1.8 Your iPhone 3G comes with a number of predefined wallpaper images.

1. **On the Home screen, tap Settings.** The Settings application appears.

2. **Tap Wallpaper.** The Wallpaper screen appears.

3. **Tap either Camera Roll or the photo album that contains the image you want to use.** iPhone 3G displays the images in the album you choose.

4. **Tap the image you want to use.** The Move and Scale screen appears, as shown in figure 1.9.

5. **Tap and drag the image so that it's positioned on the screen the way you want.**

6. **Pinch and spread your fingers over the image to set the zoom level you want.**

7. **Tap Set Wallpaper.** iPhone 3G sets the image as the wallpaper.

1.9 Use the Move and Scale screen to set the position and zoom level for the new wallpaper.

Taking a wallpaper photo with the iPhone 3G camera

For even more wallpaper fun, you can create an on-the-fly wallpaper image using the iPhone 3G camera. Here are the steps to follow:

1. **On the Home screen, tap Camera.** The Camera application appears.

2. **Line up your subject and tap the Camera button to take the picture.**

3. **Tap the Camera Roll button.** The Camera Roll photo album appears.

4. **Tap the photo you just took.** A preview of the photo appears, as shown in figure 1.10.

5. **Tap the Action button.** The Action button is the button on the left side of the menu bar. (If you don't see the menu bar, tap the screen.) iPhone 3G displays a list of actions you can perform.

6. **Tap Use As Wallpaper.** The Move and Scale screen appears.

7. **Tap and drag the image so that it's positioned on the screen the way you want.**

1.10 Tap the photo you want to use as wallpaper to see a preview of the photo.

Swap button→

8. Pinch or spread your fingers over the image to set the zoom level you want.

9. Tap Set Wallpaper. iPhone 3G sets the image as the wallpaper.

Customizing the Home Button

The Home button is the starting point for most of your iPhone 3G excursions, and it seems like the simplest of the iPhone 3G knickknacks:

● If your iPhone 3G is in Standby mode, press the Home button to display the Slide to Unlock screen.

● If your iPhone 3G is already on, press the Home button to return to the Home screen.

That's it, right? Not so fast! You can actually customize the Home button to do some useful things. No, you can't change any of the built-in behaviors that your iPhone 3G performs when you press the Home button. However, you *can* customize what your iPhone 3G does when you "double-press" the Home button. Apple actually calls this "double-clicking" the Home button, which is at least more familiar terminology, so we'll switch to that for the rest of this section.

By default, your iPhone 3G performs one of the following actions when you double-click the Home button:

- If the iPhone 3G iPod is playing, it displays the iPod Playback controls.
- If the iPhone 3G iPod is not playing, it displays the Phone application's Favorites list.

These are useful shortcuts to know, for sure, but you can customize this behavior by following these steps:

1. **On the Home screen, tap Settings.** The Settings application appears.

2. **Tap General.** The General screen appears.

3. **Tap Home Button.** The Home Button screen appears, as shown in figure 1.11.

4. **Tap the screen you want to appear when you double-click the Home button: Home, Phone Favorites, or iPod.**

5. **If you always want to see the screen you chose in step 4 when you double-click Home, change the iPod Controls setting to Off.**

1.11 Use the Home Button screen to customize Home button double-clicks.

Customizing the Keyboard

You can type on your iPhone 3G, although don't expect to pound out the prose as easily as you can on your computer. The onscreen keyboard is a bit too small for rapid and accurate typing, but it's still a far sight better than any other phone out there, mostly because the keyboard was thoughtfully designed by the folks at Apple. It even changes depending on the application you use. For example, the regular keyboard features a spacebar at the bottom. However, if you're surfing the Web with your iPhone 3G's Safari browser, the keyboard that appears when you type in the address bar does away with the spacebar. In its place you find a period (.), a slash (/), and a button that enters the characters .com. Web addresses don't use spaces so Apple replaced the spacebar with three things that commonly appear in a Web address. Nice!

Another nice innovation you get with the iPhone 3G keyboard is a feature called Auto-Capitalization. If you type a punctuation mark that indicates the end of a sentence — for example, a period (.), a question mark (?), or an exclamation mark (!) — or if you press Return to start a new paragraph, the iPhone 3G automatically activates the Shift key, because it assumes you're starting a new sentence.

On a related note, double-tapping the spacebar activates a keyboard shortcut: instead of entering two spaces, the iPhone 3G automatically enters a period (.) followed by a space. This is a welcome bit of efficiency because otherwise you'd have to tap the Number key (.?123) to display the numbers and punctuation marks, tap the period (.), and then tap the spacebar.

Genius

Typing a number or punctuation mark normally requires three taps: tapping Number (.?123), tapping the number or symbol, and then tapping ABC. Here's a faster way: press and hold the Number key to open the numeric keyboard, slide the *same* finger to the number or punctuation symbol you want, and then release the key. This types the number or symbol and returns to the regular keyboard all in one touch.

One thing the iPhone 3G keyboard doesn't seem to have is a Caps Lock feature that, when activated, enables you to type all-uppercase letters. To do this, you need to tap and hold the Shift key and then use a different finger to tap the uppercase letters. However, the iPhone 3G keyboard actually *does* have a Caps Lock feature; it's just that it's turned off by default.

To turn on Caps Lock, and to control the Auto-Capitalization and the spacebar double-tap shortcut, follow these steps:

1. **On the Home screen, tap Settings.** The Settings application appears.

2. **Tap General.** The General screen appears.

3. **Tap Keyboard.** The Keyboard screen appears, as shown in figure 1.12.

1.12 Use the Keyboard screen to customize a few keyboard settings.

18

4. **Use the Auto-Capitalization setting to turn this feature On or Off.**

5. **Use the Enable Caps Lock setting to turn this feature On or Off.**

6. **Use the "." Shortcut setting to turn this feature On or Off.**

7. **To add an international keyboard layout, tap International Keyboards to open the Keyboards screen, and then set the keyboard layout you want to add to On.**

Note
When you're using two or more keyboard layouts, the keyboard sprouts a new key to the left of the spacebar (it looks like a stylized globe). Tap that key to run through the layouts (the names of which appear briefly in the spacebar).

Resetting the iPhone 3G

If you've spent quite a bit of time in the iPhone 3G's Settings application, your phone probably doesn't look much like it did fresh out of the box. That's okay, though, because your iPhone 3G should be as individual as you are. However, if you've gone a bit *too* far with your customizations, your iPhone 3G might feel a bit alien and uncomfortable. That's okay, too, because there's an easy solution to the problem: you can erase all your customizations and revert the iPhone 3G back to its default settings.

A similar problem that comes up is when you want to sell or give your iPhone 3G to someone else. Chances are you don't want the new owner to see your data — contacts, appointments, e-mail and text messages, favorite Web sites, music, and so on — and it's unlikely the other person wants to wade through all that stuff anyway (no offense). To solve this problem, you can erase not only your custom settings, but also all of the content you've stored on the iPhone 3G.

The iPhone 3G's Reset application handles these scenarios and a few more to boot. Here's how it works:

1. **On the Home screen, tap Settings.** The Settings application appears.

2. **Tap General.** The General screen appears.

Caution
If you have any content on your iPhone 3G that isn't synced with iTunes — for example, iTunes music you've recently downloaded or an Apps Store program that you've recently installed — you lose that content if you choose Reset All Content and Settings. First sync your iPhone 3G with your computer to save your content, and then run the reset.

3. **Tap Reset.** The Reset screen appears, as shown in figure 1.13.

4. **Tap one of the following reset options:**

   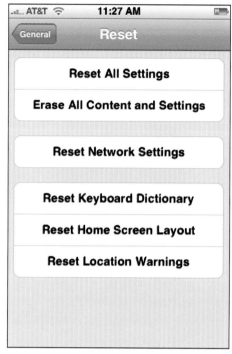

   ● **Reset All Settings.** Tap this option to reset your custom settings to the factory default settings.

   ● **Reset All Content and Settings.** Tap this option to reset your custom settings and remove any data you've stored on the iPhone 3G.

   ● **Reset Network Settings.** Tap this option to delete your Wi-Fi network settings, which is often an effective way to solve Wi-Fi problems.

   ● **Reset Keyboard Dictionary.** Tap this option to reset your keyboard dictionary. This dictionary contains a list of the keyboard suggestions that you've rejected. Tap this option to clear the dictionary and start fresh.

1.13 Use the Reset screen to reset various aspects of your iPhone 3G.

   ● **Reset Warning locations.** Tap this option to wipe out the location preferences for your applications. A location warning is the dialog box you see when you start a 6PS-aware application for the first time, and your iPhone 3G asks if the application can use your current location. You either click OK or Don't Allow, and these are the preferences you're resetting here.

5. **When the iPhone 3G asks you to confirm, tap the red button.** Note that the name of this button is the same as the reset option. For example, if you tapped the Reset All Settings option in step 4, the confirm button is called Reset All Settings. iPhone 3G resets the data.

Note Remember that the keyboard dictionary contains rejected suggestions. For example, if you type "Viv", iPhone 3G suggests "Big," instead. If you tap the "Big" suggestion to reject it and keep "Viv," the word "Big" is added to the keyboard dictionary.

How Can I Get More Out of My iPhone 3G's Phone Feature?

The iPhone 3G is chock full of great applications that enable you to surf the Web, send and receive e-mail messages, listen to music, take photos, organize your contacts, schedule appointments, and much, much more. These features put the "smart" into the iPhone 3G's status as a smartphone, but let's not forget the "phone" part! So while you're probably familiar with the basic steps required to make and answer calls, the iPhone 3G's powerful phone component is loaded with amazing features that can make the cell phone portion of your life easier, more convenient, and more efficient. This chapter takes you through these features.

Working with Outgoing Calls

You can do much more with your iPhone 3G than just make a call the old-fashioned way — by dialing the phone number. There are speedy shortcuts you can take, and even settings to alter the way your outgoing calls look on the receiver's phone.

Making calls quickly

The iPhone 3G has a seemingly endless number of methods you can use to make a call. It's nice to have the variety, but in this have-your-people-call-my-people world, the big question is not how many ways can you make a call, but how *fast* can you make a call? Here are our favorite iPhone 3G speed-calling techniques:

- **Favorites list.** This list acts as a kind of speed dial for the iPhone 3G because you use it to store the phone numbers you call most often, and you have space to add your top 20 numbers. To call someone in your Favorites list, double-click the Home button to leap immediately to the Favorites screen, and then tap the number you want to call. How to manage your Favorites is covered later in this chapter.

- **Visual Voicemail.** If you're checking your voicemail messages (from the Home screen, tap Phone, and then tap Voicemail) and you want to return someone's call, tap the message and then tap Call Back.

- **Text message.** If someone types a phone number in a text message, iPhone 3G handily converts that number into a kind of link: The number appears in blue, underlined text, much like a link on a Web page, as shown in figure 2.1. Tap the phone number to call that number. You can also use a similar technique to call numbers embedded in Web pages (see Chapter 3) and e-mail messages (see Chapter 4).

2.1 Your iPhone 3G is kind enough to convert a text message phone number into a link that you can tap to call the number.

⬤ **Recent numbers.** The Recent Calls list (from the Home screen, tap Phone and then tap Recents) shows your recent phone activity: the calls you've made, the calls you've received, and the calls you've missed. Recent Calls is great because it enables you to quickly redial someone you've had recent contact with. Just tap the call and away you go. (If you want to return a missed call, tap Missed and then tap the call.) To call the person using a different phone number, tap the More Info icon (the arrow) to the right of the name or number, and then tap the phone number you want to use to make the call.

Genius   If your Recent Calls list is populated with names or numbers that you know you won't ever call back, you should clear the list and start fresh. In the Recent Calls screen, tap Clear and then tap Clear All Recents.

Configuring your iPhone 3G to not show Your caller ID

When you use your iPhone 3G to call someone, and the called phone supports Caller ID, your number and often your name appear. If you'd rather hide your identity for some reason, you can configure your iPhone 3G to not show your caller ID:

1. **On the Home screen, tap Settings.** The Settings application appears.

2. **Tap Phone.** The Phone screen appears.

3. **Tap Show My Caller ID.** The Show My Caller ID screen appears.

4. **Tap the Show My Caller ID On/Off button to change this setting to Off, as shown in figure 2.2.** Your iPhone 3G disables the Caller ID feature.

2.2 Change the Show My Caller ID setting to Off to prevent your iPhone 3G from displaying your caller ID when you phone someone.

25

Handling Incoming Calls

When a call comes into your iPhone 3G, you answer it, right? What could be simpler? You'd be surprised. Your iPhone 3G gives you quite a few options to dealing with that call, aside from just answering it. After all, you don't want to talk to everyone all the time, do you?

Silencing an incoming call

When you're in a situation where the ringing of a cell phone is inappropriate, bothersome, or just plain rude, you, as a good cell phone citizen, practice "celliquette" (cell etiquette) and turn off your ringer. (On your iPhone 3G, flick the Silent/Ring switch on the left side panel to the silent position.) However, we are merely human and so we all forget to turn off our phone's ringer once in a while. Hey, it happens.

Your job in that situation is to grab your phone and answer it as quickly as possible. However, what if you're in a situation where answering the call is bad form? Or what if you'd prefer to delay answering the call until you can leave the room or get out of earshot? That's a stickier cell wicket, for sure, but the iPhone 3G designers have been there and they've come up with a simple solution: press either the Sleep/Wake button on the phone's top panel, or either volume button on the left side panel. Either way, your iPhone 3G stops ringing (and vibrating). The ringing is still going on (your caller hears it on her end), so you've still got the usual four rings to answer the call should you decide to.

Genius If you don't want someone to know you are ignoring his or her call, just silence the ring. The caller will still hear the standard four rings before the voicemail and be none the wiser that you just didn't pick up your phone.

Sending an incoming call directly to voicemail

Sometimes you just don't want to talk to someone. Whether this person is your significant other calling to complain, a friend who never seems to have anything to say and just talks in circles for ten minutes, or if you're just indisposed at the moment, you might prefer to ignore the call.

That's not a problem on your iPhone 3G:

- If the phone isn't locked, tap the red Decline button on the touchscreen.
- If you're using the earbuds you just need to squeeze and hold the microphone/clicker for two seconds.
- Press the Sleep/Wake button twice in quick succession.

Any of these methods sends the call directly to voicemail.

Caution

If you ignore a call, as with any phone, the caller will know that you've ignored the call when voicemail kicks in before the normal four rings.

Turning off the iPhone 3G call waiting feature

If you're already on a call and another call comes in, your iPhone 3G springs into action and displays the person's name or number as well as three options (described in detail earlier in this chapter): Ignore, Hold Call + Answer, and End Call + Answer. This is part of your iPhone 3G's Call Waiting feature, and it's great if you're expecting an important call or if you want to add the caller to a conference call that you've set up.

However, the rest of the time you might just find it annoying and intrusive (and anyone who you put on hold or hang up on to take the new call probably finds it rude and insulting). In that case, you can turn off Call Waiting by following these steps:

1. **On the Home screen, tap Settings.** The Settings application appears.

2. **Tap Phone.** The Phone screen appears.

3. **Tap Call Waiting.** The Call Waiting screen appears.

4. **Tap the Call Waiting On/Off button to change this setting to Off, as shown in figure 2.3.** Your iPhone 3G disables the Call Waiting feature.

2.3 Change the Call Waiting setting to Off to disable your iPhone 3G's Call Waiting feature.

Forwarding iPhone 3G calls to another number

What do you do about incoming calls if you can't use your iPhone 3G for a while? For example, if you're going on a flight, you must either turn off your iPhone 3G or put it into Airplane mode, as described in this chapter, so incoming calls won't go through. Similarly, if you have to return your iPhone 3G to Apple for repairs or battery replacement, the phone won't be available if anyone tries to call you.

For these and other situations where your iPhone 3G can't accept incoming calls, you can work around the problem by having your calls forwarded to another number, such as your work number or your home number. Here's how it's done:

1. **On the Home screen, tap Settings.** The Settings application appears.

2. **Tap Phone.** The Phone screen appears.

3. **Tap Call Forwarding.** The Call Forwarding screen appears.

4. **Tap the Call Forwarding On/Off button to change this setting to On.** Your iPhone 3G displays the Forwarding To screen.

5. **Tap the phone number to use for the forwarded calls.**

6. **Tap Call Forwarding to return to the Call Forwarding screen.** Figure 2.4 shows the Call Forwarding screen set up to forward your iPhone 3G calls.

2.4 Activate Call Forwarding to have your iPhone 3G calls forwarded to another number.

Juggling Multiple Calls and Conference Calls

We all juggle multiple tasks and duties these days, so it's not surprising that sometimes that involves juggling multiple phone calls:

- You might need to call two separate people on a related issue, and then switch back and forth between the callers as the negotiations (or whatever) progress.

- You might already be on a call and another call comes in from a person you need to speak to, so you put the initial person on hold, deal with the new caller, and then return to the first person.

- You might need to speak to two separate people at the same time on the same phone call — in other words, a conference call.

In the real world, juggling multiple calls and setting up conference calls often requires a special phone or a fancy phone system. In the iPhone 3G world, however, these things are a snap. In fact, the way the iPhone 3G juggles multiple calls really is something spectacular. Jumping back and forth between calls is simple; putting someone on hold to answer an incoming call is a piece of cake; and creating a conference call from incoming or outgoing calls is criminally easy.

When you're on an initial call, your iPhone 3G displays the Call Options screen shown in figure 2.5. To make another call, tap Add Call and then use the Phone application to place your second call.

Once the second call goes through, the Call Options screen changes: The top of the screen shows the first caller's name or number, with HOLD beside it, and below that you see the name or number of the second call and the duration of that call. Figure 2.6 shows the new screen layout. To switch to the person on hold, tap the Swap button. iPhone 3G puts the second caller on hold and returns you to the first caller. Congratulations: You now have two separate calls going at once!

2.5 When you're on a call, your iPhone 3G displays these call options.

29

If you're already on the phone and another call comes in, your iPhone 3G displays the number (and the name, if the caller is in your Contacts list), and gives you three ways to handle the call (see figure 2.7):

- **Ignore.** Tap this option to send the incoming call directly to voicemail.

- **Hold Call + Answer.** Tap this option to put the first call on hold and answer the incoming call. You're working with two separate calls again in this scenario, so you can tap Swap to switch between the callers.

- **End Call + Answer.** Tap this option to drop the first call and answer the incoming call.

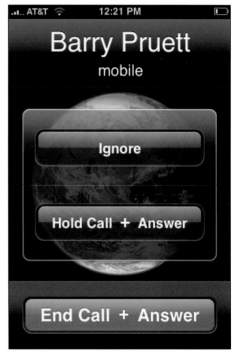

2.6 The iPhone 3G Call Options screen with two separate phone calls on the go.

2.7 The iPhone 3G displays this screen if a call comes in while you are on another call.

If you have two calls on the go, you might prefer that all three of you be able to talk to each other in a conference call. Easier done than said: tap the Merge option and iPhone 3G combines everyone into a single conference call and displays Conference at the top of the Call Options screen. Click the More Info arrow and iPhone 3G displays the participants' names or numbers in the Conference screen, as shown in figure 2.8.

From here, there are a few methods you can use to manage your conference call:

- To speak with one of the callers privately, tap the green Private key next to that person's name or number you want to talk with. This places you in a one-on-one call with that person and places the other caller on hold.

- To drop someone from the conference call, tap the red phone icon to the left of the person's name or number, and then tap End Call to confirm. iPhone 3G drops the caller and you resume a private call with the other caller.

- To add someone else to the conference call, tap Back to return to the Call Options screen, tap Add Call, and then make the call. Once the call goes through, tap Merge Calls.

- To add an incoming caller to the conference call, tap Hold Call + Answer. Once you're connected, tap Merge Calls.

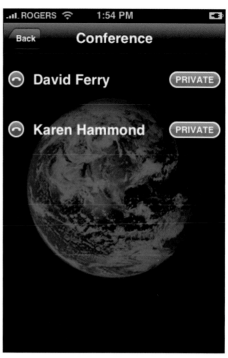

2.8 When you merge two phone calls, the participants' names or numbers appear in the Conference screen.

Clearly, juggling multiple calls on a phone has never been easier. The iPhone 3G does a remarkable job of organizing the calls and giving you an admirably easy process to follow to swap calls, add or drop calls, and combine calls in conference.

Caution

You can hold a conference call with up to five people at once by repeating the steps outlined for conference calls. However, remember, conference calls use up your minutes faster — two callers use them up twice as fast, three callers use them up three times as fast, and so on — so you may want to be judicious when using this feature.

Using Other iPhone 3G Features While On a Call

We live in a multitasking age, and your iPhone 3G can multitask with the best of them. For example, suppose you're on a call and the other person needs someone's phone number or e-mail address. No sweat: with your iPhone 3G, you can switch to the Contacts list, get the info you seek, and recite it to your caller, all without interrupting the call. In fact, you can switch to any iPhone 3G application during the call: you can look up information on the Web using Safari; set up an appointment; look up a map location; send an e-mail; check your text messages; even crank up a tune using the iPod!

Here are the steps to follow:

1. **Initiate the phone call.**

2. **Tap the Speaker icon.** This ensures that you can still converse with the caller while using the other application.

3. **When you need to use a different application, press the Home button.** iPhone 3G displays the Home screen, but you remain connected to the caller. iPhone 3G displays a bar across the top of the screen (below the status bar) that says Touch to return to call, as shown in figure 2.9.

4. **Tap the icon of the application you want to use.**

5. **When you complete your chores in the application, tap the Touch to return to call bar.** iPhone 3G returns you to the Call Options screen.

6. **Tap the Speaker icon to turn it off and then continue with the call.**

2.9 When you're on a call and you switch to another application, you see "Touch to return to call" below the status bar.

Managing Your Favorites List

The iPhone 3G's Favorites list is great for making quick calls because you can often get someone on the horn in just two finger gestures (double-click the Home button and then tap the number). Of course, this only works if the numbers you call most often appear on your Favorites list. Fortunately, your iPhone 3G gives you lots of different ways to populate the list. Here are the easiest methods to use:

- In the Favorites list, click + to open the All Contacts screen and then tap the person you want to add. If that person has multiple phone numbers, tap the number you want to use as a favorite.

Note This is a good place to remind you that the Favorites list isn't a list of people, it's a list of *numbers*. That's why the list shows both the person's name and the type of phone number (work, home, mobile, and so on).

- In the Recent Calls list, tap the More Info icon to the right of the call from (or to) the person you want to add and then tap Add to Favorites. If the person has multiple phone numbers, tap the number you want to use as the favorite. iPhone 3G adds a star beside the phone number to remind you that it's a favorite.

- In Visual Voicemail, tap the More Info icon beside a message and then tap Add to Favorites.

- In the Contacts list (tap Contacts in the Home screen), tap the person you want to add and then tap Add to Favorites. If the person has multiple phone numbers, tap the number you want to use as the favorite. iPhone 3G adds a star beside the phone number to remind you that it's a favorite.

You can add up to 20 numbers in the Favorites list, but the iPhone 3G screen only shows eight numbers at a time. This means that if you want to call someone who doesn't appear in the initial screen, then you need to scroll down to bring that number into view. Therefore, your Favorites list is most efficient when the people you call most often appear in the first eight numbers. Your iPhone 3G adds each new number to the bottom of the Favorites list, so chances are that at least some of your favorite numbers aren't showing up in the top eight. Follow these steps to fix that:

1. **In the Favorites list, tap Edit.** iPhone 3G displays delete icons to the left of each favorite and drag icons to the right, as shown in figure 2.10.

2. **If you want to get rid of a favorite, tap its Delete icon, tap Remove, and then tap Edit to return to Edit mode.**

3. **To move a favorite to a new location, tap and drag the icon up or down until the favorite is where you want it, and then release the icon.**

4. **Click Done.**

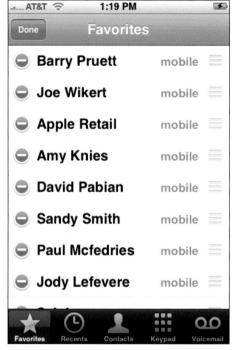

2.10 In Edit mode, the Favorites list shows delete icons on the left and drag icons on the right.

Working with Contacts from the Phone Application

As you learn later in this book, you can enter all your contacts into your iPhone 3G — either by syncing with your computer or by adding them on the fly. You do all that using your iPhone 3G's Contacts application, but the Phone application also has a couple of useful tricks you can use. You can convert a phone number into a contact and even assign specific ringtones to various contacts.

Converting a phone number into a contact

Your iPhone 3G is at its most efficient when the numbers you call are part of your Contacts list, because not only can you add contacts to the Favorites list for quick, speed-dial-like access, but also because you can use the index (the letters A, B, C, and so on that run down the right side of the Contacts list) and a few finger flicks to rapidly find and tap the person you want to chinwag with.

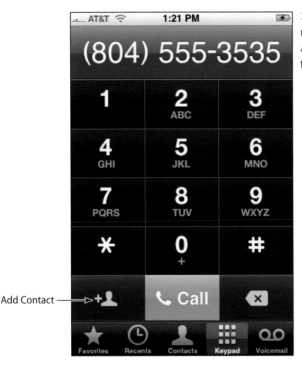

2.11 After you enter a phone number using the Keypad, tap the Add Contact icon to create a contact for the number.

Add Contact ⟶

We talk about ways to add contacts in Chapter 7. For now, here's a quick way to add a contact right from your iPhone 3G's phone keypad:

1. **In the Home screen, tap Phone.** The Phone application appears.

2. **In the menu bar, tap Keypad.** The phone keypad appears.

3. **Type the phone number of a person you want to add as a contact.**

4. **Tap the Add Contact icon, to the left of the green Call button, as shown in figure 2.11.**

5. **Tap Create New Contact.** The New Contact screen appears.

6. **Fill in the other contact info, as needed.**

7. **Tap Save.** Your iPhone 3G adds the new contact and returns you to the Keypad screen.

8. **Tap Call to proceed with the phone call.**

Note

The phone number you're dialing might be an alternative number of an existing contact. For example, you may already have set up the contact with a home number, but now you're dialing that person's cell number. In that case, tap the Add Contact icon and then tap Add to Existing Contact. Use the Contacts list to tap the contact, choose the phone type (such as mobile), and then tap Save.

Assigning a ringtone to a contact

Your iPhone 3G has a default ringtone that it plays whenever a call comes in. You grab your phone, check the name (if the person's in your Contacts list) or the number, and then decide whether to answer or let the call go to voicemail. However, what if you're busy or concentrating on something and you'd prefer not to break off just to check the incoming call? Wouldn't it be nice to *know* whether the call is important?

Your iPhone 3G smartphone isn't quite smart enough to know that, but you can help it along by assigning different ringtones to different contacts. There are a couple of ringtone routes you can take:

- **You can assign a different ringtone to each of the people who call you most often.** That way, you can know exactly who's calling you just by hearing the ringtone.

- **You can assign a single different ringtone to all of the people who you consider important.** That way, when you hear that ringtone, you know that it's okay to interrupt whatever you're doing; if you hear the regular ringtone, just keep working (or whatever).

Here are the steps to follow to assign a ringtone to a contact:

1. **On the Home screen, tap Contacts.** Alternatively, if you're currently in the Phone application, tap Contacts in the menu bar.

2. **Tap the contact you want to work with.** The contact's Info screen appears.

3. **Tap the Ringtone setting.** The iPhone 3G displays the Ringtones screen.

4. **Tap the ringtone you want to use.** iPhone 3G plays a preview. If that's not the tone you want, tap another until you find the right one.

5. **Tap Info.** iPhone 3G returns you to the contact's Info screen with the new ringtone selected.

Working with Voicemail

Voicemail is a great thing because even if you miss a call, you never really miss a call (providing the caller leaves a message, of course). But, what if you're trying to navigate a voicemail system from your iPhone 3G or even setting up your own voicemail? What then?

Entering an extension or navigating a voicemail system

When people see the iPhone 3G for the first time, the lack of a physical keypad is the feature that really throws them for a loop. With most other phones, you can start dialing immediately just by poking numbers on the keypad, but with the iPhone 3G you have to display the Keypad screen and *then* tap the number.

Once you're on a call, however, a second keypad conundrum surfaces: How do you enter numbers? For example, when the lady with the nice voice says, "If you know the extension of the person you wish to reach, please enter it now," how exactly do you do that?

The secret is in the call options that your iPhone 3G displays when you're connected. Those options include a Keypad icon, and tapping that icon displays the Keypad screen shown in figure 2.12 which, as you can see, is similar to the one you use for dialing numbers directly. When you're prompted for a number

2.12 When you're on a call, tap the Keypad icon to see this Keypad screen.

or a symbol such as star (*) or pound (#), tap Keypad, tap the number or symbol, and then tap Hide Keypad.

That method works fine if you just need to enter a single extension or symbol, but it quickly gets comically inefficient when you're navigating a voicemail program or some other system with a seemingly endless series of prompts. You bring the phone to your ear to hear the prompt, and then bring the phone in front of you to tap the required code on the keypad, then return the phone to your ear for the next prompt, and the cycle goes on and on. It's a surefire recipe for tennis elbow (iPhone elbow?).

Note

An extra problem is created by one of the iPhone 3G's coolest features. When you bring the phone to your ear while on a call, the iPhone 3G senses the new orientation and automatically blanks the screen to save battery power. How smart is *that*! However, when you bring the phone in front of you to enter something on the keypad, you have to wait a second or two before the screen comes back to life.

Here's a much better method for navigating voicemail and similar multiprompt systems:

1. **Initiate the phone call.** iPhone 3G displays the call options.
2. **Tap the Speaker icon.** This tells the iPhone 3G to output the sounds through the speaker on the bottom panel.
3. **Tap the Keypad icon to display the on-call keypad.**
4. **While keeping the phone in front of you, listen to and respond to the system's prompts.**
5. **When the prompts are done, tap Hide Keypad to return to the Call Options screen.**
6. **Tap the Speaker icon to turn it off and then continue with the call.**

Working with your voicemail

The voicemail program is a basic part of any phone, but the iPhone 3G's Visual Voicemail feature is anything but basic. The iPhone 3G takes advantage of its beautiful screen to organize your voice-mails visually, and does a stunning job of it, as you can see in figure 2.13. Unlike most other phones, you don't have to go through an automated process to get your voicemail. In fact, you don't have to listen to them in order. You can skip the one sent by that friend who never quite seems to say anything and listen to the one from a coworker telling you it's free pretzel day and he's saving a spot in line for you. The iPhone 3G itself stores your messages, so you don't have to be within your cellular provider's service area to hear your messages.

To get to the Voicemail screen, tap the Phone icon on the Home screen, and then tap the Voicemail icon in the menu bar.

Setting up voicemail

The first time you land on the Voicemail screen, your iPhone 3G prompts for a numeric password. You use the password to check your messages from phones other than your iPhone 3G.

The Voicemail application then asks you to enter a greeting message, which is the message that folks hear when their calls get dumped (or sent!) to voicemail. You have two choices (see figure 2.14):

- **Default.** Tap this option to use the iPhone 3G default message, which is a female voice saying the call has been forwarded to a voice message system.

- **Custom.** Tap this option, tap Record, and then record your message. When you're done, tap Stop, and then tap Save.

2.13 The iPhone 3G makes your voicemail a visual affair.

2.14 For your iPhone 3G voicemail greeting, you can choose either the phone's default message or you can record a custom message.

If you want to change your custom message later, follow these steps:

1. **In the Home screen, tap Phone.** The Phone application appears.

2. **Tap Voicemail.** The Voicemail screen appears.

3. **Tap Greeting.** The Greeting screen appears.

4. **Tap Custom.** You only need to do this if you currently have the Default option selected.

5. **Tap Record.** Use the iPhone 3G microphone to record your message.

6. **Tap Stop when you finish the message.**

7. **To review the greeting to see if it's worthy, tap Play.**

8. **If you like what you hear, tap Save.** If you don't like the message, tap Cancel and then repeat steps 5 to 8.

Accessing your messages

How do you know if you have any voicemail messages waiting for you? Tap the Home screen's Phone icon and then eyeball the Voicemail icon in the menu bar. If the red circle in the upper-right corner is empty, then no messages for you! Otherwise, you see a number inside the circle that tells you how many messages are waiting for you (see figure 2.15).

2.15 If you've got voicemail, the Voicemail icon tells you how many messages are standing by.

To access your messages, tap the Voicemail icon. Your iPhone 3G whisks you to the Voicemail screen and displays a list of your messages. (This is why Apple calls this *Visual* Voicemail: You can actually see your messages!) As you can see in figure 2.16, the entries show the name of the caller (if that caller is in your Contacts list) or number of the caller (if your iPhone 3G has no idea who called). To the right of the entry you see the time the message was left (if the message came in today) or the date the message was left (if the message came in earlier). A blue dot to the left of the entry indicates a message that you haven't heard yet.

You can scroll through your messages by sliding your finger up or down the screen. When you find the message you want to work with, you can use any of the following techniques:

- **Listen to a message.** Double-tap the message. While the message is playing, you see a playback bar, and a white ball moves along this bar to indicate the current position within a message. To fast-forward or rewind the message, drag the ball forward or backward along the bar.

Note If you've got a few messages to slog through, you might want to turn on your iPhone 3G speaker to listen to your messages. To do this, tap the Speaker button at the top right of the Voicemail screen.

See information about the caller. Tap the blue More Info arrow next to the sender's name or number to access the More Info screen for the message. This shows you the time and date that the message was sent followed by the sender info if the sender is in your Contacts. The number from which the message was sent is in blue. If the sender is not in your Contacts, you see the number and the city and state that the phone is registered in. You are also provided with a Create New Contact and Add to Existing Contact button. For both cases, you have the option of returning the call or sending a text from this screen.

2.16 The Voicemail screen lists all your saved voicemail messages.

 Return the call. Tap the message and then tap Call Back.

 Delete the message. Tap the message and then tap Delete. If you delete a message by accident, tap Deleted Messages (it's at the bottom of the Voicemail screen), tap the message you want to restore, and then tap Undelete.

Caution

There are many other places on the iPhone 3G where you are prompted to confirm a delete. In the Visual Voicemail, a single touch does it. Although you can always undelete a message, be careful around the Delete key in this program.

Note

To access your voicemail from another phone, call your number. When the greeting begins, press *****, then enter your numeric password, and then press **#**. Follow the instructions given.

41

Switching Your iPhone 3G to Airplane Mode

When you board a flight, aviation regulations in most countries are super strict about cell phones: no calls in and no calls out. In fact, most of those regulations ban wireless signals of *any* kind, which means your iPhone 3G is a real hazard to sensitive airline equipment because it also transmits Wi-Fi and Bluetooth signals, even if there are no Wi-Fi receivers or Bluetooth devices within 30,000 feet of your current position.

Your pilot or friendly flight attendant will suggest that everyone simply turn off their phones. Sure, that does the job, but darn it you've got an *iPhone 3G*, which means there are plenty of things you can do outside of its wireless capabilities: listen to music or an audiobook, watch a show, view photos, and much more.

So how do you reconcile the no-wireless-and-that-means-you regulations with the iPhone 3G's multitude of wireless-free applications? You put your iPhone 3G into a special state called Airplane mode. This mode turns off the transceivers — the internal components that transmit and receive wireless signals — for the iPhone 3G's phone, Wi-Fi, and Bluetooth features. With your iPhone 3G now safely in compliance of federal aviation regulations, you're free to use any application that doesn't rely on wireless transmissions.

Follow these steps to activate Airplane mode:

1. **On the Home screen, tap Settings.** The Settings application appears.

2. **Tap the Airplane Mode On/Off switch to turn this setting On, as shown in figure 2.17.** Notice that while Airplane mode is on, an Airplane icon appears in the status bar in place of the signal strength and network icons.

Note

If a flight attendant sees you playing around with your iPhone 3G, he or she may ask you to confirm that the phone is off. (One obviously iPhone-savvy attendant even asked me if my phone was in Airplane mode.) Showing the Airplane icon should be sufficient.

Airplane mode icon→

2.17 When your iPhone 3G is in Airplane mode, an Airplane icon appears in the status bar.

Connecting Your iPhone 3G with a Bluetooth Headset

Your iPhone 3G is configured to use a wireless technology called Bluetooth, which enables you to make wireless connections to other Bluetooth-friendly devices. Most Macs come with Bluetooth built in, and they can use it to connect to a wide range of Bluetooth devices, including a mouse, keyboard, cell phone, PDA, printer, digital camera, and even another Mac. Your iPhone 3G isn't so versatile, however. All you can do on the Bluetooth front is connect to a headset, which at least lets you listen to phone conversations, music, and movies without wires and without disturbing your neighbors.

In theory, connecting Bluetooth devices should be criminally easy: You turn on each device's Bluetooth feature — in Bluetooth jargon, you make the device *discoverable* — bring them within 33 feet of each other, and they connect without further ado. In practice, however, there's usually at least a bit of further ado (and sometimes plenty of it). This usually takes one or both of the following forms:

43

● **Making your device discoverable**. Unlike Wi-Fi devices that broadcast their signals constantly, most Bluetooth devices only broadcast their availability when you say so. This makes sense in many cases because you usually only want to connect a Bluetooth component such as a headset with a single device. By controlling when the device is discoverable, you ensure that it works only with the device you want it to.

● **Pairing iPhone 3G and the device.** As a security precaution, many Bluetooth devices need to be *paired* with another device before the connection is established. In most cases, the pairing is accomplished by typing a multidigit *passkey* — your iPhone 3G calls it a PIN — that you must then type into the Bluetooth device (assuming, of course, that it has some kind of keypad). In the case of a headset, the device comes with a default passkey that you must enter into your iPhone 3G to set up the pairing.

Making your iPhone 3G discoverable

So your first order of Bluetooth business is to ensure that your iPhone 3G is discoverable by activating the Bluetooth feature. This feature is on by default with new iPhone 3Gs, so check for that first: On the status bar, look for the Bluetooth logo to the left of the battery status icon, as shown in figure 2.18.

Bluetooth icon

2.18 If your iPhone 3G is discoverable, you see the Bluetooth icon in the status bar.

A Bit of Bluetooth Background

You're probably familiar with Wi-Fi, the standard that enables you to perform networking chores without the usual network cables. Bluetooth is similar in that it enables you to exchange data between two devices without any kind of physical connection between them. Bluetooth uses radio frequencies to set up a communications link between the devices. That link is an example of an *ad hoc wireless network* — a network without a central connection point — only in this case the network that Bluetooth creates is called a *piconet*.

Bluetooth is a short-distance networking technology, with a maximum range of about 33 feet (10 meters). The Bluetooth name comes from Harald Bluetooth, a tenth-century Danish king who united the provinces of Denmark under a single crown, the same way that, theoretically, Bluetooth will unite the world of portable wireless devices under a single standard. Why name a modern technology after an obscure Danish king? Here's a clue: Two of the most important companies backing the Bluetooth standard — Ericsson and Nokia — are Scandinavian.

If you don't see the Bluetooth icon, follow these steps to turn on Bluetooth and make your iPhone 3G discoverable:

1. **On the Home screen, tap Settings.** The Settings application appears.

2. **Tap General.** The General screen appears.

3. **Tap Bluetooth.** The Bluetooth screen appears.

4. **Tap the Bluetooth On/Off button to change the setting to On, as shown in figure 2.19.**

Pairing your iPhone 3G with a Bluetooth headset

If you want to listen to music, headphones are a great way to go because the sound is often better than with the built-in iPhone 3G speakers, and no one else around is subjected to Weezer

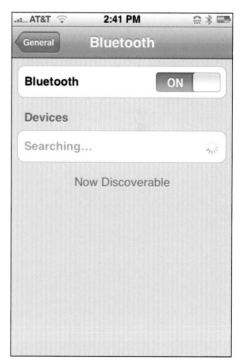

2.19 Use the Bluetooth screen to make your iPhone 3G discoverable.

45

at top volume. Similarly, if you want to conduct a hands-free call, a headset (a combination of head-phones for listening and a microphone for talking) makes life easier because you can put the phone down and make all the hand gestures you want (providing you aren't driving, of course). Add Bluetooth into the mix, and you've got an easy and wireless audio solution for your iPhone 3G.

Follow these general steps to pair your iPhone 3G with a Bluetooth headset:

1. **On the Home screen, tap Settings.** The Settings application appears.

2. **Tap General.** The General screen appears.

3. **Tap Bluetooth.** The Bluetooth screen appears.

4. **If the headset has a separate switch or button that makes the device discoverable, turn on that switch or press that button.** Wait until you see the correct headset name appear in the Bluetooth screen, as shown in figure 2.20.

5. **Tap the name of the Bluetooth headset.** Your iPhone 3G should pair with the headset automatically and you see Paired in the Bluetooth screen, as shown in figure 2.21; you can skip the rest of these steps. Otherwise you see the Enter PIN screen.

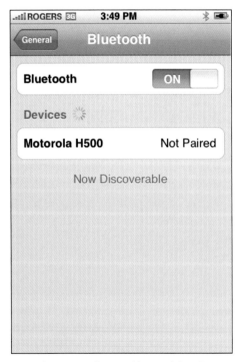

2.20 When you make your Bluetooth headset discoverable, the device appears in the Bluetooth screen.

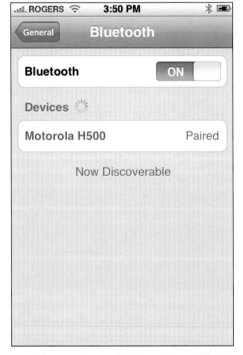

2.21 When you have paired your iPhone 3G with the Bluetooth headset, you see Paired beside the device in the Bluetooth screen.

6. **Tap the headset's passkey in the PIN box.** See the headset documentation to get the passkey (it's often 0000).

7. **Tap Connect.** Your iPhone 3G pairs with the headset and returns you to the Bluetooth screen, where you now see Paired beside the headset name.

8. **Click Quit and the headset is ready to use.**

Unpairing your iPhone 3G from a Bluetooth headset

When you no longer plan to use a Bluetooth headset for a long period of time, you should unpair it from your iPhone 3G. Follow these steps:

1. **On the Home screen, tap Settings.** The Settings application appears.

2. **Tap General.** The General screen appears.

3. **Tap Bluetooth.** The Bluetooth screen appears.

4. **Tap the name of the Bluetooth headset.**

5. **Tap Unpair.** Your iPhone 3G unpairs the headset.

The Apple Bluetooth Earpiece

Apple's Bluetooth earpiece is compact and lightweight and is sold exclusively for the iPhone. It retails for $130 and comes with a charging cradle that charges the earpiece and the iPhone simultaneously. The good people at Apple also provide a little bonus. When you charge the two together they automatically pair. You don't have to go through the steps outlined in this section. Now *that's* simple.

The earpiece has only one button, and this button connects it to the iPhone. You see a small icon at the top of your iPhone's screen when the two are connected. You operate the iPhone 3G as normal, only you use the earpiece to hear and speak. To answer a call, press the earpiece button. Press it again to end the call.

Turning Off Data Roaming

Data roaming is an often convenient cell phone feature that enables you to make calls — and, with your iPhone 3G, surf the Web, check and send e-mail, and exchange text messages — when you're outside of your provider's normal coverage area. The downside is that roaming charges are almost always eye-popping expensive, and we're often talking several dollars per *minute*, depending on where you are and what type of service you're using. Not good!

Unfortunately, if you have your iPhone 3G's Data Roaming feature turned on, you may incur massive roaming charges even if you never use your phone! That's because your iPhone 3G still performs background checks for things like incoming e-mail messages and text messages, so a week in some far-off land could cost you hundreds of dollars without even using your phone.

To avoid this insanity, turn off your iPhone 3G's Data Roaming feature when you don't need it. Follow these steps:

1. **On the Home screen, tap Settings.** The Settings application appears.

2. **Tap General.** The General screen appears.

3. **Tap Network.** The Network screen appears.

4. **Tap the Data Roaming On/Off button to change this setting to Off, as shown in figure 2.22.**

2.22 Use the Network screen to change the Data Roaming setting to Off.

How Can I Make the Most of Web Surfing with My iPhone 3G?

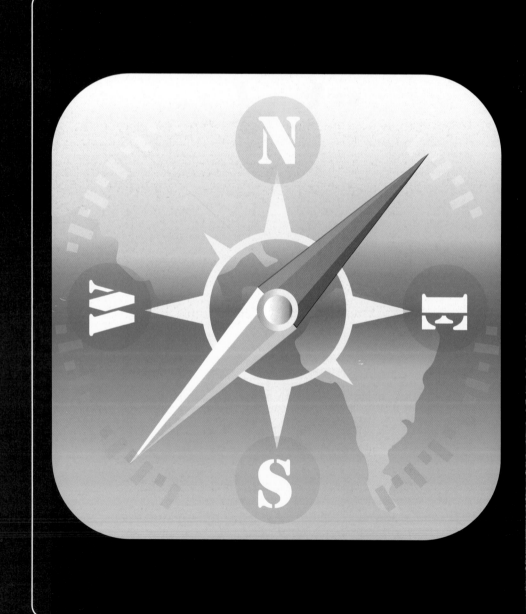

One of the most popular modern pastimes is Web surfing, and now you can surf sites even when you're out and about thanks to your iPhone 3G's large screen and support for speedy networks such as 3G and Wi-Fi. You perform these surfin' safaris using, appropriately enough, the Mobile Safari Web browser. This is a trimmed-down version of Safari, one of the world's best browsers, and out of the box it's easy to use and intuitive. However, Mobile Safari offers quite a few options and features, many of which are hidden in obscure nooks and crannies of the iPhone 3G interface. If you think your surfing activities could be faster, more efficient, more productive, or more secure, this chapter shows you a bunch of techniques that can help.

Understanding Internet Access Networks

To get on the Web, your iPhone 3G must first connect to a network that offers Internet access. To make this easy and seamless, your iPhone 3G can work with not just a single network, not just two networks (like the original iPhone), but *three* network types:

- **Wi-Fi**. Short for wireless fidelity, and also called 802.11 by the geeks and AirPort by Apple types, you use your iPhone 3G's built-in Wi-Fi antenna to connect to a wireless network that's within range. This usually means within about 115 feet, but some public networks — also called *Wi-Fi hot spots* — boost their signals to offer a greater range. Because most wireless networks are connected to high-speed Internet connections, Wi-Fi is by far your best bet for an Internet connection. You get fast downloads and you don't use up data in whatever Internet connection plan you have with your cellular provider. Plus you get the added bonus of being able to make and receive phone calls while you're online. As long as a Wi-Fi network is within range and you can connect to that network, your iPhone 3G always defaults to using Wi-Fi for Internet access.

Genius

Just about any place you see people set up with their laptops has Wi-Fi that you can use. Another easy way to find Wi-Fi near you is to open the Google Maps application on your iPhone 3G and type your city and **wifi** into the search box. This gives you a map with pushpins representing Wi-Fi hotspots near where you are.

- **3G**. Short for Third Generation, the 3G cellular network is what gives your iPhone 3G its name. It's slower than Wi-Fi, but Wi-Fi requires you to be in a hot spot to use it. The 3G network is a cellular network, so as long as you're in a 3G coverage area, you can access the Internet from anywhere, even a moving car. 3G is currently available in over 280 U.S. metropolitan areas and AT&T plans to have it up and running in 350 by the end of 2008. Most other countries offer widespread 3G coverage, so you won't often find yourself out of 3G service. As a major bonus for cell phone addicts everywhere, 3G lets you access the Internet *and* talk on the phone at the same time, something the original iPhone's EDGE network couldn't do. Plus, the 3G download speeds are anywhere from 2 to 2.5 times as fast as the notoriously pokey EDGE downloads, so you won't grow old waiting for a Web site to open. If your iPhone 3G has no Wi-Fi hot spot in range, it automatically switches to the 3G network, assuming you're in a coverage area.

- **EDGE.** This is short for Enhanced Data rates for GSM (Global System for Mobile communi-cation) Evolution, an absurdly grandiose name for a rightfully maligned cellular network technology. Why the bad press for EDGE? Because, in a word, it's *slow*. Paint dries faster than most Web sites download over an EDGE connection. Even worse, EDGE is strictly a monotasking system: If you're on the phone, you can't surf the Web; if you're surfing the Web, you can't send or receive calls. So why bother with EDGE at all? Mostly because although 3G is widespread, it doesn't have as much coverage as EDGE does. So if you don't have a Wi-Fi network nearby, and you're not in a 3G coverage area, your iPhone 3G drops down into EDGE mode so you can at least get a signal.

Connecting to a Wi-Fi Network

Connections to the cellular network are automatic and occur behind the scenes. As soon as you switch on your iPhone 3G, it checks for a 3G signal and, if it finds one, it connects to the network and displays the 3G icon in the status bar, as well as the connection strength (the more bars the bet-ter). If your current area doesn't do the 3G thing, your iPhone 3G tries to connect to an EDGE net-work instead. If that works, you see the E icon in the status bar (plus the usual signal strength bars). If none of that works, you see No Signal, so you might as well go home.

Making your first connection

Things aren't immediately automatic when it comes to Wi-Fi connections, at least not at first. As soon as you try to access something on the Internet — a Web site, your e-mail, a Google Map, or whatever — your iPhone 3G scours the surrounding airwaves for Wi-Fi network sig-nals. If you've never connected to a Wi-Fi net-work, or if you're in an area that doesn't have any Wi-Fi networks that you've used in the past, you see the Select a Wi-Fi Network dialog box, as shown in figure 3.1.

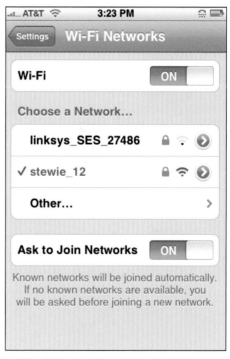

3.1 If you're just starting out on the Wi-Fi trail, or if you're blazing a new Wi-Fi trail, your iPhone 3G displays a list of nearby networks.

This dialog box displays a list of the Wi-Fi networks that are within range. For each network, you get three tidbits of data:

- **Network name.** This is the name that the administrator has assigned to the network. If you're in a coffee shop or similar public hot spot and you want to use that network, look for the name of the shop (or a variation on the name).

- **Password-protected.** If a Wi-Fi network displays a lock icon, it means the network is protected by a password, and you need to know that password to make the connection.

- **Signal strength.** This icon gives you a rough idea of how strong the wireless signals are. The stronger the signal (the more bars you see, the better the signal), the more likely you are to get a fast and reliable connection.

Follow these steps to connect to a Wi-Fi network:

1. **Tap the network you want to use.** If the network is protected by a password, your iPhone 3G prompts you to enter the password, as shown in figure 3.2.

2. **Use the keyboard to tap the password.**

3. **Tap Join.** The iPhone 3G connects to the network and adds the Wi-Fi network signal strength icon to the status bar.

To connect to a commercial Wi-Fi operation — such as those you find in airports, hotels, and convention centers, for example — you almost always have to take one more step. In most cases, the network prompts you for your name and credit card data so you can be charged for accessing the network. If you're not prompted right away, you will be as soon as you try to access a Web site or check your e-mail. Enter your information and then enjoy the Internet in all of its Wi-Fi glory.

3.2 If the Wi-Fi network is secured with a password, use this screen to enter it.

Caution Because the password box shows dots instead of the actual text for added security, this is no place to demonstrate your iPhone 3G speed-typing prowess. Slow and steady wins the password typing race (or something).

Note If you're not fortunate enough to be near Wi-Fi you can use, you can still access the Internet via the cellular connection. Your iPhone 3G tries to use the 3G network, but if that's a no go it uses the dreaded EDGE network.

Connecting to known networks

If the Wi-Fi network is one that you use all the time — for example, your home or office network — the good news is that your iPhone 3G remembers any network that you connect to. As soon as a known network comes within range, your iPhone 3G makes the connection without so much as a peep. Thanks!

Stopping the incessant Wi-Fi network prompts

The Select a Wi-Fi Network dialog box is a handy convenience if you're not sure whether a Wi-Fi network is available. However, as you move around town, you may find that dialog box popping up all over the place as new Wi-Fi networks come within range. One solution is to wear your finger down to the bone with all the constant tapping of the Cancel button, but there's a better way: just tell your iPhone 3G to shut up already with the Wi-Fi prompting. Here's how:

1. **On the Home screen, tap Settings.** The Settings screen appears.

2. **Tap Wi-Fi.** iPhone 3G opens the Wi-Fi Networks screen.

3. **Tap the Ask to Join Networks switch to the Off position, as shown in figure 3.3.** Your iPhone 3G no longer prompts you with nearby networks. Whew!

Okay, we hear you ask, if I'm no longer seeing the prompts, how do I connect to a Wi-Fi network if I don't even know it's there? That's a good question, and here's a good answer:

1. **On the Home screen, tap Settings.** Your iPhone 3G displays the Settings screen.

2. **Tap Wi-Fi.** The Wi-Fi Networks screen appears, and the Choose a Network list shows you the available Wi-Fi networks.

3. **Tap the network you want to use.** If the network is protected by a password, your iPhone 3G prompts you to enter the password.

4. **Use the keyboard to tap the password.**

5. **Tap Join.** The iPhone 3G connects to the network and adds the Wi-Fi network signal strength icon to the status bar.

Connecting to a hidden Wi-Fi network

Each Wi-Fi network has a network name — often called the Service Set Identifier, or SSID — that identifies the network to Wi-Fi-friendly devices such as your iPhone 3G. By default, most Wi-Fi networks broadcast the network name so that you can see the network and connect to it. However, some Wi-Fi networks disable network name broadcasting as a security precaution. The idea here is that if an unauthorized user can't see the network, he or she can't attempt to connect to it. (However, some devices can pick up the network name when authorized computers connect to the network, so this is not a foolproof security measure.)

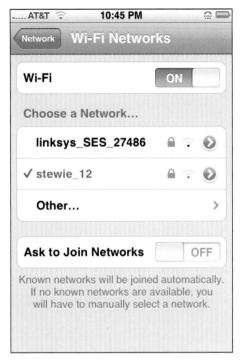

3.3 Toggle the Ask to Join Networks switch to Off to put a gag on the network prompts.

You can still connect to a hidden Wi-Fi network by entering the connection settings by hand. You need to know the network name, the network's security type and encryption type, and the network's password. Here are the steps to follow:

1. **On the Home screen, tap Settings to open the Settings screen.**

2. **Tap Wi-Fi.** You see the Wi-Fi Networks screen.

3. **Tap Other.** Your iPhone 3G displays the Other Network screen, as shown in figure 3.4.

4. **Use the Name text box to type the network name.**

5. **Tap Security to open the Security screen.**

6. **Tap the type of security used by the Wi-Fi network: WEP, WPA, WPA2, WPA Enterprise, WPA2 Enterprise, or None.**

7. **Tap Other Network to return to the Other Network screen.** If you chose WEP, WPA, WPA2, WPA Enterprise, or WPA2 Enterprise, your iPhone 3G prompts you to enter the password.

8. **Use the keyboard to tap the password.**

9. **Tap Join.** The iPhone 3G connects to the network and adds the Wi-Fi network signal strength icon to the status bar.

3.4 Use the Other Network screen to connect to a hidden Wi-Fi network.

Turning off the Wi-Fi antenna to save power

Your iPhone 3G's Wi-Fi antenna is constantly on the lookout for nearby Wi-Fi networks. That's useful because it means you always have an up-to-date list of networks to check out, but it does take its toll on the iPhone 3G battery. If you know you won't be using Wi-Fi for a while, you can save some battery juice for more important pursuits by turning off your iPhone 3G's Wi-Fi antenna. Here's how:

1. **On the Home screen, tap Settings.** The Settings screen appears.

2. **Tap Wi-Fi.** The Wi-Fi Networks screen appears.

3. **Tap the Wi-Fi switch to the Off position.** Your iPhone 3G disconnects from your current Wi-Fi network and hides the Choose a Networks list.

When you're ready to resume your Wi-Fi duties, return to the Wi-Fi Networks screen and tap the Wi-Fi switch to the On position.

Touchscreen Tips for Web Sites

The touchscreen operates much the same way in Safari as it does in the other iPhone 3G applications. You can use the touchscreen to scroll pages, zoom in and out, click links, fill in forms, enter
addresses, and more. The screen is remarkably fluid in its motion and its response to your touch is
neither hyperactive nor sluggish. It actually makes surfing the Web a pleasure, which isn't something you can say about most smartphones.

To make it even more pleasurable, here's a little collection of touchscreen tips that ought to make
your Web excursions even easier:

- **Precision zooming.** Zooming on the iPhone 3G is straightforward: to zoom in, spread
  two fingers apart; to zoom out, pinch two fingers together. However, when you zoom in
  on a Web page, it's almost always because you want to zoom in on *something*. It might be
  an image, a link, a text box, or just a section of text. To ensure that your target ends up in
  the middle of the zoomed page, pinch your thumb and forefinger together on the screen
  as if you are pinching the target you want to zoom in on. Spread your thumb and forefinger apart to zoom in.

- **The old pan-and-zoom.** Another useful technique for getting a target in the middle of a
  zoomed page is to zoom and pan at the same time. That is, as you spread (or pinch) your
  fingers, you also move them up, down, left, or right to pan the page at the same time. This
  takes a bit of practice, and often the iPhone 3G only allows you to pan either horizontally
  or vertically (not both), but it's still a useful trick.

- **Double-tap.** A quick way to zoom in on a page that has various sections is to double-tap
  on the specific section — it could be an image, a paragraph, a table, or a column of text —
  that you want magnified. Your iPhone 3G zooms the section to fill the width of the screen.
  Double-tap again to return the page to the regular view.

Note

The double-tap-to-zoom trick only works on pages that have identifiable sections. If
a page is just a wall of text, you can double-tap until the cows come home (that's a
long time) and nothing much happens.

- **One tap to the top.** If you're reading a particularly long-winded Web page and you're near
  the bottom, you may have quite a long way to scroll if you need to head back to the top to
  get at the address bar or tap the Search icon. Save the wear-and-tear on your flicking finger! Instead, tap the status bar; Safari immediately transports you to the top of the page.

Tap and hold to see where a link takes you. You "click" a link in a Web page by tapping it with your finger. In a regular Web browser, you can see where a link takes you by hovering the mouse pointer over the link and checking out the link address in the status bar. That doesn't work in your iPhone 3G, of course, but you can still find out the address of a link before tapping it. Hold your finger on the link for a few seconds and Safari displays a pop-up balloon that shows the link text and, more importantly, the link address, as shown in figure 3.5. If the link looks legit, lift your finger to complete the tap; if you decide not to go there, you can bail out of the tap by moving your finger off the link, and then lifting it off the screen.

Use the portrait view to navigate a long page. When you rotate your iPhone 3G 90 degrees, the touchscreen switches to landscape view, which gives you a wider view of the page. Return the iPhone 3G to its upright position, and you return to portrait view. If you have a long way to scroll in a page, first use the portrait view to scroll down then switch to the landscape view to increase the text size. We've found that scrolling in the portrait view goes much faster than in landscape.

Two-fingered frame scrolling. Some Web sites are organized using a technique called *frames*, where the overall site takes up the browser window, but some of the site's pages appear in a separate rectangular area — called a *frame* — usually with its own scroll bar. In such sites, you may find that the usual one-fingered scroll technique only scrolls the entire browser window, not the content within the frame. To scroll the frame stuff, you must use *two* fingers to do the scrolling. Weird!

3.5 Hold your finger on a link to see the link address.

- **Getting a larger keyboard.** The onscreen keyboard appears when you tap into a box that allows typing. We've noticed, however, that the keyboard you get in landscape view uses noticeably larger keys than the one you see in portrait view. For the fumble-fingered among us, larger keys are a must, so always rotate the iPhone 3G into landscape mode to enter text. Note that you need to make the switch to landscape *before* you tap inside the text box; otherwise, once the keyboard is onscreen, you can't switch the view.

- **Save typing with standard Web addresses.** Most Web sites have addresses that start with http://www. and end with .com/. Safari on your iPhone 3G knows this, and it uses this otherwise unremarkable fact to save you a ton of typing. If you type just a single block of text into the address bar — it could be a single word such as wiley or two or more words combined into one, such as wordspy — and then tap Go, Safari automatically adds http://www. to the front and .com/ at the end. So wiley becomes http://www.wiley.com/ and wordspy becomes http://www.wordspy.com/.

Filling in Online Forms

Many Web pages include forms where you fill in some data and then submit the form, which sends the data off to some server for processing. Filling in these forms in your iPhone 3G's Safari browser is mostly straightforward:

- **Text box.** Tap inside the text box to display the touchscreen keyboard, tap out your text, and then tap Done.

- **Text area.** Tap inside the text area, and then use the keyboard to tap your text. Most text areas allow multiline entries, so you can tap Return to start a new line. When you finish, tap Done.

- **Check box.** Tap the check box to toggle the check mark on and off.

- **Radio button.** Tap a radio button to activate it.

- **Command button.** Tap the button to make it do its thing (usually submit the form).

Many online forms consist of a bunch of text boxes or text areas. If the idea of performing the tap-type-Done cycle over and over isn't appealing to you, fear not. Your iPhone 3G's Safari browser offers an easier method:

1. **Tap inside the first text box or text area.** The keyboard appears.

2. **Tap to type the text you want to enter.** Above the keyboard, notice the Previous and Next buttons, as shown in figure 3.6.

3. **Tap Next to move to the next text box or text area.** If you need to return to a text box, tap Previous instead.

4. **Repeat steps 2 and 3 to fill in the text boxes.**

5. **Tap Done.** Safari returns you to the page.

We haven't yet talked about selection lists, and that's because your iPhone 3G's browser handles them in an interesting way. When you tap a list, Safari displays the list items in a separate box, as shown in figure 3.7. Tap the item you want to select. As with text boxes and text areas, if the form has multiple lists, you see the Previous and Next buttons, which you can tap to navigate from one list to another. After you make all your selections, tap Done to return to the page.

3.6 If the form contains multiple text boxes or text areas, you can use the Previous and Next buttons to navigate them.

3.7 Tap a list to see its items in a separate box for easier selecting.

Using Bookmarks for Faster Surfing

Although you've seen that your iPhone 3G's Safari browser offers a few tricks to ease the pain of typing Web page addresses, it's still slower and quite a bit more cumbersome than a full-size, physical keyboard, which lets even inexpert typists rattle off addresses lickety-split. All the more reason that you should embrace bookmarks with all your heart. After all, a bookmark lets you jump to a Web page with precisely no typing; just a tap or three and you're there.

Adding bookmarks by hand

You probably want to get your iPhone 3G bookmarks off to a flying start by copying a bunch of existing bookmarks from your Mac or Windows PC. That's a good idea, and we show you how to do that in Chapter 5.

But even if you've done the sync and now have a large collection of bookmarks at your beck and call, it doesn't mean your iPhone 3G bookmark collection is complete. After all, you might find something interesting while you're surfing with the iPhone 3G. If you think you'll want to pay that site another visit down the road, you can create a new bookmark right on the iPhone 3G. Here are the steps to follow:

1. **On the iPhone 3G.** Use Safari to navigate to the site you want to save.
2. **Tap the + button in the menu bar.**
3. **Tap Add Bookmark.** This opens the Add Bookmark screen, as shown in figure 3.8.
4. **Tap into the top box and type a name for the site that helps you remember it.** This name is what you see when you scroll through your bookmarks.
5. **Tap Bookmarks.** This displays a list of your bookmark folders.

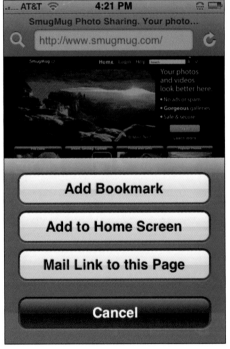

3.8 Use the Add Bookmark screen to specify the bookmark name and location.

6. **Tap the folder you want to use to store the bookmark.** Safari returns you to the Add Bookmark screen.

7. **Tap Save.** Safari saves the bookmark.

Note

Syncing bookmarks is a two-way street, which means that any site you bookmark in your iPhone 3G gets added to your desktop Safari (or Internet Explorer) the next time you sync.

Getting Firefox bookmarks into your iPhone 3G

iTunes bookmark syncing only works with Safari and Internet Explorer. So are you out of luck if your entire Web life is bookmarked in Firefox? Nope. Fortunately Firefox has a feature that lets you export your bookmarks to a file. You can then import those bookmarks to Safari or Internet Explorer, and then sync with your iPhone 3G. It's a bit of a winding road, we know, but it's better than starting from scratch. Here are the details:

1. **Crank up Firefox and start the export procedure like so:**

   ● **Firefox 3 (Mac).** Choose Bookmarks ⇨ Organize Bookmarks (or press Shift+⌘+B) to open the Library. Click the Import and Backup button, and then click Export HTML.

   ● **Firefox 3 (Windows).** Choose Bookmarks ⇨ Organize Bookmarks (or press Ctrl+Shift+B) to open the Library. Click the Import and Backup button, and then click Export HTML.

   ● **Firefox 2 (Windows).** Choose Bookmarks ⇨ Organize Bookmarks to open the Bookmarks Manager, and then choose File ⇨ Export.

2. **In the Export Bookmarks File dialog box, choose a location for the file, and then click Save.** Firefox saves its bookmarks to a file named bookmarks.html.

3. **Import the Firefox bookmarks file to your browser of choice:**

   ● **Safari.** Choose File ⇨ Import Bookmarks, locate and click the bookmarks.html file, and then click Import.

   ● **Internet Explorer.** In version 7, click the Add to Favorites button (or press Alt+Z) and then click Import and Export; in version 6, choose File ⇨ Import and Export. In the Import/Export Wizard, click Next, click Import Favorites, and then follow the wizard's instructions to import the bookmarks.html file.

4. **Connect your iPhone 3G to your computer.** iTunes opens, connects to the iPhone 3G, and syncs the bookmarks, which now include your Firebox bookmarks.

Managing your bookmarks

Once you have a few bookmarks stashed away in the bookmarks list, you may need to perform a few housekeeping chores from time to time, including changing a bookmark's name, address, or folder; reordering bookmarks or folders; or getting rid of bookmarks that have worn out their welcome.

Before you can do any of this, you need to get the Bookmarks list into Edit mode by following these steps:

Bookmarks icon

1. **In Safari, tap the Bookmarks icon in the menu bar (see figure 3.9).** Safari opens the Bookmarks list.

3.9 Tap the Bookmark icon to display the Bookmarks list.

2. **If the bookmark you want to mess with is located in a particular folder, tap to open that folder.** For example, if you've synced with Safari, then you should have a folder named Bookmarks Bar, which includes all the bookmarks and folders that you've added to the Bookmarks Bar in your desktop version of Safari.

3. **Tap Edit.** Your iPhone 3G switches the Bookmarks list to Edit mode, as shown in figure 3.10.

   With Edit mode on the go, you're free to toil away at your bookmarks. Here are the techniques to master:

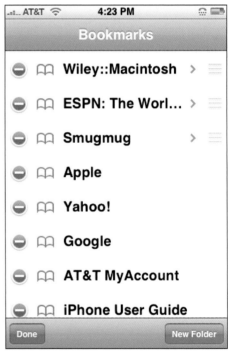

3.10 With the Bookmark list in Edit mode, you can edit, rearrange, and delete bookmarks to your heart's content.

- **Edit bookmark info.** Tap the bookmark to fire up the Edit Bookmark screen. From here, you can edit the bookmark name, change the bookmark address, and change the bookmark folder. When you're done, tap the name of the current bookmark folder in the top-left corner of the screen.

- **Change the bookmark order.** Use the drag icon on the right to tap-and-drag a bookmark to a new position in the list. Ideally, you should move your favorite bookmarks near the top of the list for easiest access.

- **Add a bookmark folder.** Tap New Folder to launch the Edit Folder screen, then tap a folder title and select a location. Feel free to use bookmark folders at will because they're a great way to keep your bookmarks neat and tidy (if you're into that kind of thing).

- **Delete a bookmark.** No use for a particular bookmark? No problem. Tap the Delete icon to the left of the bookmark, and then tap the Delete button that appears.

When the dust settles and your bookmark chores are done for the day, tap Done to get out of Edit mode.

Retracing your steps with the handy History list

Bookmarking a Web site is a good idea if that site contains interesting or fun content that you want to revisit in the future. Sometimes, however, you may not realize that a site had useful data until a day or two later. Similarly, you might like a site's stuff, but decide against bookmarking it, only to regret that decision down the road. You could waste a big chunk of your day trying to track down the site, but then you may run into Murphy's Web Browsing Law: A cool site that you forget to bookmark is never found again.

Fortunately, your iPhone 3G has your back. As you navigate the nooks and crannies of the Web, iPhone 3G keeps track of where you go and stores the name and address of each page in the History list. The iPhone 3G's limited memory means that it can't store tons of sites, but it might have what you're looking for. Here's how to use it:

1. **In Safari, tap the Bookmarks icon in the menu bar.** Safari opens the Bookmarks list.

2. **Tap the folder names that appear in the upper-left corner of the screen until you get to the Bookmarks screen.**

3. **Tap History.** Safari opens the History screen, as shown in figure 3.11. The screen shows the sites you've visited today at the top, followed by a list of previous surfing dates.

3.11 Safari stores your recent browsing past in the History list.

65

4. **If you visited the site you're looking for on a previous day, tap the day.** Safari displays a list of the sites you visited on that day.

5. **Tap the site you want to revisit.** Safari loads the site.

Maintaining your privacy by deleting the History list

Your iPhone 3G's History list of sites you've recently surfed is a great feature when you need it, and it's an innocuous feature when you don't. However, there are times when the History list is just plain uncool. For example, suppose you shop online to get a nice gift for your spouse's birthday. If he or she also uses your iPhone 3G, your surprise might get ruined if the purchase page accidentally shows up in the History list. Similarly, if you visit a private corporate site, a financial site, or any other site you wouldn't want others to see, the History list might betray you.

And sometimes unsavory sites can end up in your History list by accident. For example, you might tap a legitimate-looking link in a Web page or e-mail message, only to end up in some dark, dank Net neighborhood. Of course, you high-tailed it out of there right away with a quick tap of Safari's Back button, but that nasty site is now lurking in your History.

Whether you've got sites on the History list that you wouldn't want anyone to see, or if you just find the idea of your iPhone 3G tracking your movements on the Web to be a bit sinister, follow these steps to wipe out the History list:

1. **In Safari, tap the Bookmarks icon in the menu bar.** Safari opens the Bookmarks list.

2. **Tap the folder names that appear in the upper-left corner of the screen until you get to the Bookmarks screen.**

3. **Tap History.** Safari opens the History screen.

4. **Tap Clear.** Safari prompts you to confirm.

5. **Tap Clear History.** Safari deletes every site from the History list.

Genius   Here's another way to clear the History, and it might be faster if you're not currently working in Safari. In the Home screen, tap Settings, tap Safari, and then tap Clear History. When your iPhone 3G asks you to confirm, tap Clear History.

Opening and Managing Multiple Browser Pages

When you're perusing Web pages, what do you do when you're on a page that you want to keep reading, but you also need to leap over to another page for something? On your computer's Web browser, you probably open another tab, use that tab to open the other page, and then switch back to the first page when you finish. It's an essential Web browsing technique, but can it be done with your iPhone 3G's Safari browser?

Well, Mobile Safari may not have tabs, but it has the next best thing: *pages*. With this feature, you can open a second browser "window" and load a different page into it, and then it's a quick tap and flick to switch between them. And you're not restricted to a meager two pages, no sir. Your iPhone 3G lets you open up to eight — count 'em, *eight* — pages, so you can throw some wild Web page parties.

Here are the steps to follow to open and load multiple pages:

1. **In Safari, tap the Pages icon in the menu bar (see figure 3.12).** Safari displays a thumbnail version of the current page.

   Pages icon

   3.12 Tap the Pages icon to create a new page.

2. **Tap New Page.** Safari opens a blank page using the full screen.

3. **Load a Web site into the new page.** You can do this by selecting a bookmark, typing an address, or whatever.

4. **Repeat Steps 1 to 3 to load as many pages as you need.** As you add pages, Safari keeps track of how many are open and displays the number in the Pages icon, as shown in figure 3.13.

   3.13 The Pages icon tells you how many pages you've got on the go.

Note Some Web page links are configured to automatically open the page in a new window, so you might see a new page being created when you tap a link. Also, if you add a Web Clip to your Home screen (as described in Chapter 1), tapping the icon opens the Web Clip in a new Safari page.

Once you have two or more pages fired up, here are a couple of techniques you can use to impress your friends:

- **To switch to another page.** Tap the Pages icon to get to the thumbnail view (see figure 3.14), flick right or left to bring the page into view, and then tap the page.

- **When you no longer need a page.** Tap the Pages icon, flick right or left to bring the page into view, and then tap the X in the upper-left corner. Safari trashes the page without a whimper of protest.

3.14 Tap the Pages icon to see thumbnail versions of your open pages.

Note

Below the page thumbnails you see several dots, one for each open page. If you know what order you loaded the pages (and if you've honed your precision tapping skills), a quicker way to switch to a page is to tap its dot.

Changing the Default Search Engine

When you tap the Search icon at the top of the Safari screen, your iPhone 3G loads the Address Bar screen and places the cursor inside the Search box so that you can type your search text and then run the search. The button you tap to launch the search is named Google, which is appropriate as Google is the iPhone 3G's default search engine. We all love Google, of course, but if you have something against it, for some reason, you can switch to using Yahoo! as your search engine of choice. Here's how:

1. **In the Home screen, tap Settings.** Your iPhone 3G opens the Settings screen.

2. **Tap Safari.** The Safari screen appears.

3. **Tap Search Engine.** Your iPhone 3G opens the Search Engine screen, as shown in figure 3.15.

4. **Tap Yahoo!.** Your iPhone 3G now uses Yahoo! as the default search engine. In the Search screen, you can now enter your search text and tap the Yahoo! button.

Viewing an RSS Feed

Some Web sites remain relatively static over time, so you only need to check in every once in a while to see if anything's new. Other sites change content regularly, such as once a day or once a week, so you know in advance when to check for new material. However, there are the more verbose sites — particularly blogs — where the content changes fre-

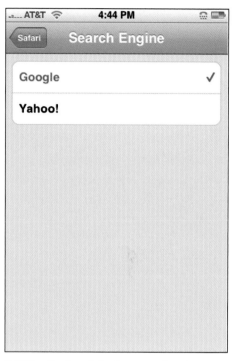

3.15 Use the Search Engine screen to set your iPhone 3G's default search service.

quently, although not regularly. For these sites, keeping up with new content can be time consuming, and it's criminally easy to miss new information. (Murphy's Blog Reading Law: You always miss the post that everyone's talking about.)

To solve this problem, tons of Web sites now maintain RSS feeds (RSS stands for Real Simple Syndication). A *feed* is a special file that contains the most recent information added to the site. The bad news is that your iPhone 3G's Safari browser doesn't give you any way to subscribe to a site's feed, like you can with desktop Safari or Internet Explorer. The good news is that your iPhone 3G can use a Web-based RSS reader application (http://reader.mac.com/) that can interpret a site's RSS feed and display the feed in the comfy confines of Safari.

Here's how it works:

1. **In Safari, navigate to a Web page that you know has an RSS feed.**

2. **Pan and zoom the page until you find the link to the RSS feed.** The link is often accompanied by (or consists entirely of) an icon that indentifies it as leading to a feed. Look for an XML icon, an RSS icon, or an orange feed icon, as shown in figure 3.16.

3. **Tap the link.** Safari loads the RSS file into the reader.mac.com feed reader application, as shown in figure 3.17.

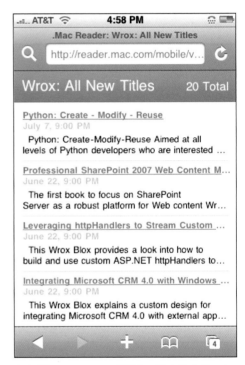

3.16 Most feed links are identified by a standard feed icon.

3.17 When you tap an RSS feed link, Safari loads the RSS file into the reader.mac.com feed reader application.

Dialing a Phone Number from a Web Page

Your iPhone 3G's membership in the smartphone club is fully confirmed with this next feature. A common chore when you're surfing business or retail sites is to hunt down a phone number for a person, a department, Customer Service, Technical Support, or whatever. In a regular browser, you note the phone number, head for the nearest phone, and then dial. Hah, the iPhone 3G laughs at all that extra work! Why? Because when its Safari browser comes upon a phone number in a Web page, Safari conveniently converts that number into a link, as shown in figure 3.18. Tap the number, tap Call in the dialog box that pops up, and you're immediately switched to the Phone application, which dials the number for you. Sweet!

Setting the Web Browser Security Options

It's a jungle out there in cyberspace, with nasty things lurking in the digital weeds. The folks at Apple are well aware of these dangers, of course, so they've clothed your iPhone 3G in protective gear to help keep the bad guys at bay. Safari, in particular, has four layers of security:

3.18 Your iPhone 3G smartly converts a Web page phone number into a link that you can tap to call the number.

- **JavaScript.** This is a programming language that Web site developers commonly use to add features to their pages. However, programmers who have succumbed to the dark side of The Force can use JavaScript for nefarious ends. Your iPhone 3G comes with JavaScript support turned on, but you can turn it off if you're heading into an area of the Web where you don't feel safe. However, many sites won't work without JavaScript, so we don't recommend turning it off full time.

- **Plug-ins.** This is a kind of mini-application that extends the features of the Safari browser. Most browser plug-ins add functionality such as support for playing certain types of video and audio files, and they're perfectly safe. However, there are malicious plug-ins that can hijack the browser and direct Safari to evil sites. If you're worried about this, you can turn off Safari's plug-in support.

● **Pop-up blocking.** Pop-up ads (and their sneakier cousins, the pop-under ads) are annoy-ing at the best of times, but they really get in the way on the iPhone 3G because the pop-up not only creates a new Safari page, but it immediately switches to that page. So now you have to tap the Pages icon, delete the pop-up page, and then (if you already had two or more pages running) tap the page that generated the pop-up. Boo! So you can thank your preferred deity that not only does your iPhone 3G come with a pop-up blocker that stops these pop-up pests, but it's turned on by default, to boot. However, there are sites that use pop-ups for legitimate reasons: media players, login pages, important site announcements, and so on. For those sites to work properly you may need to turn off the pop-up blocker temporarily.

● **Cookies.** These are small text files that many sites store on the iPhone 3G, and they use those files to store information about your browsing session. The most common example is a shopping cart, where your selections and amounts are stored in a cookie. However, for every benign cookie there's at least one not-so-nice cookie used by a third-party advertiser to track your movements and display ads supposedly targeted to your tastes. Yuck. By default, your iPhone 3G doesn't accept third-party cookies, so that's a good thing. However, you can configure Safari to accept every cookie that comes its way or no cookies at all (neither of which we recommend).

Follow these steps to customize your iPhone 3G's Web security options:

1. **In the Home screen, tap Settings.** The Settings screen slides in.

2. **Tap Safari.** Your iPhone 3G displays the Safari screen, as shown in figure 3.19.

3. **Tap the JavaScript setting to toggle JavaScript support On and Off.**

4. **Tap the Plug-Ins setting to toggle plug-in support On and Off.**

5. **Tap the Block Pop-ups setting to tog-gle pop-up blocking On and Off.**

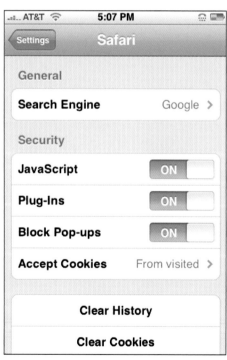

3.19 Use the Safari screen to set your iPhone 3G's Web security settings.

6. **To configure the cookies that Safari allows, tap Accept Cookies, tap the setting you want — None, From visited, or Always — and then tap Safari.** The From visited setting (the default) means that Safari only accepts cookies directly from the sites you surf to; it spits out any cookies from third-party sites such as advertisers.

7. **If you want to get rid of all the cookies that have been stored on your iPhone 3G, tap Clear Cookies and, when you're asked to confirm, tap Clear Cookies.** It's a good idea to clear cookies if you're having trouble accessing a site or if you suspect some unwanted cookies have been stored on your iPhone 3G (for example, if you surfed for a while with Accept Cookies set to Always.)

How Do I Maximize E-mail on My iPhone 3G?

E-mail has been called the "killer app" of the Internet, and it certainly deserves that title. Yes, chat and instant messaging are popular; social networks such as MySpace, Facebook, and LinkedIn get lots of press; and microblogging sites such as Twitter appeal to a certain type of person. However, while not everyone uses these services, it's safe to say that almost *everyone* uses e-mail. You probably use e-mail all day, particularly when you're on the go with your iPhone 3G in tow, so learning a few useful and efficient e-mail techniques can make your day a bit easier and save you time for more important pursuits.

Managing Your iPhone 3G E-mail Accounts

Your iPhone 3G comes with the Mail application, which is a radically slimmed-down version of the Mail application that comes with Mac machines. Mail in iPhone may be a pale shadow of its OS X cousin, but that doesn't mean it's a lightweight, far from it. It has a few features and settings that make it ideal for your traveling e-mail show. First, however, you've got to set up your iPhone 3G with one or more e-mail accounts.

Adding an account by hand

The Mail application on your iPhone 3G is most useful when it's set up to use an e-mail account that you also use on your computer. That way, when you're on the road or out on the town, you can check your messages and rest assured that you won't miss anything important (or even anything unimportant, for that matter). This is most easily done by syncing an existing e-mail account between your computer and your iPhone 3G, and we show you how that's done in Chapter 5.

Note For some accounts, you need to be careful that your iPhone 3G doesn't delete incoming messages from the server before you have a chance to download them to your computer. We show you how to set this up later in this chapter.

However, you might also prefer to have an e-mail account that's iPhone 3G-only. For example, if you join an iPhone mailing list, you might prefer to have those message sent to just your iPhone. That's a darn good idea, but it does mean that you have to set up the account on the iPhone 3G itself, which, as you soon see, requires a fair amount of tapping.

Genius You might think you're pretty smart and try to avoid the often excessive tapping required to enter a new e-mail account into your iPhone 3G by creating the account in your computer's e-mail program and then syncing with your iPhone 3G. That works, but there's a hitch: You *must* leave the new account in your e-mail program. If you delete it or disable it, iTunes also deletes the account from the iPhone 3G.

How you create an account on your iPhone 3G with the sweat of your own brow depends on the type of account you have. First, there are the five e-mail services that your iPhone 3G recognizes:

- **Microsoft Exchange.** Your iPhone 3G supports accounts on Exchange servers, which are common in large organizations like corporations or schools. Exchange uses a central server to store messages, and you usually work with your messages on the server, not your iPhone 3G. However, one of the great new features in the iPhone 3G is support for Exchange ActiveSync, which automatically keeps your phone and your account on the server synchronized. We discuss the ActiveSync settings later in this chapter.

- **MobileMe.** This is Apple's replacement for its venerable .Mac online service. We give you the details in Chapter 9.

- **Google Gmail.** This is a Web-based e-mail service run by Google.

- **Yahoo! Mail.** This is a Web-based e-mail service run by Yahoo!.

- **AOL.** This is a Web-based e-mail service run by AOL.

Your iPhone 3G knows how to connect with these services, so to set up any of these e-mail accounts you only need to know the address and the account password.

Otherwise, your iPhone 3G Mail application supports the following e-mail account types:

- **POP.** Short for Post Office Protocol, this is the most popular type of account. Its main characteristic for our purposes is that incoming messages are only stored temporarily on the provider's mail server. When you connect to the server, the messages are down-loaded to iPhone 3G and removed from the server. In other words, your messages (including copies of messages you send) are stored locally on your iPhone 3G. The advantage here is that you don't need to be online to read your e-mail. Once it's downloaded to your iPhone 3G, you can read it or delete it at your leisure.

- **IMAP.** Short for Internet Message Access Protocol, this type of account is most often used with Web-based e-mail services. It's the opposite of POP (sort of) because all your incoming messages, as well as copies of messages you send, remain on the server. In this case, when Mail works with an IMAP account, it connects to the server and works with the messages on the server itself, not on your iPhone 3G (although it *looks* like you're work-ing with the messages locally). The advantage here is that you can access the messages from multiple devices and multiple locations, but you must be connected to the Internet to work with your messages.

Your network administrator or your e-mail service provider can let you know what type of e-mail account you have. They also provide you with the information you need to set up the account. This includes your e-mail address; the username and password you use to check for new messages (and perhaps also the security information you need to specify to send messages); the host name of the incoming mail server (typically something like mail.*provider*.com, where *provider*.com is the domain name of the provider); and the host name of the outgoing mail server (typically either mail.*provider*.com or smtp.*provider*.com).

With your account information at the ready, follow these steps to forge a brand-new account:

1. **On the Home screen, tap Settings.** Your iPhone 3G opens the Settings screen.

2. **Tap Mail, Contacts, Calendars.** The Mail, Contacts, Calendars screen appears.

3. **Tap Add Account.** This opens the Add Account screen, as shown in figure 4.1.

4. **You have two ways to proceed:**

   • **If you're adding an account for Microsoft Exchange, MobileMe, Google Gmail, Yahoo! Mail, or AOL, tap the corresponding logo.** In the account information screen that appears, enter your name, e-mail address, password, and an account description. Tap Save and you're done!

4.1 Use the Add Account screen to choose the type of e-mail account you want to add.

   • **If you're adding another account type, tap Other and continue with step 5.**

5. **Use the Name, Address, and Description text boxes to type the corresponding account information, and then tap Save.**

6. **Tap the type of account you're adding: IMAP or POP.**

7. **In the Incoming Mail Server section, use the Host Name text box to enter the host name of your provider's incoming mail server, as well as your username and password.**

8. **In the Outgoing Mail Server (SMTP) section, use the Host Name text box to enter the host name of your provider's outgoing (SMTP) mail server.** If your provider requires a username and password to send messages, enter those as well.

9. **Tap Save.** Your iPhone 3G verifies the account info and then returns you to the Mail settings screen with the account added to the Accounts list.

Specifying the default account

If you've added two or more e-mail accounts to your iPhone 3G, Mail specifies one of them as the default account. This means that Mail uses this account when you send a new message, when you reply to a message, and when you forward a message. The default account is usually the first account you add to your iPhone 3G. However, you can change this by following these steps:

1. **On the Home screen, tap Settings.** The Settings screen appears.

2. **Tap Mail, Contacts, Calendars.** Your iPhone 3G displays the Mail, Contacts, Calendars screen.

3. **At the bottom of the screen, tap Default Account.** This opens the Default Account screen, which displays a list of your accounts. The current default account is shown with a check mark beside it, as shown in figure 4.2.

4. **Tap the account you want to use as the default.** Your iPhone 3G places a check mark beside the account.

5. **Tap Mail to return to the Mail settings screen.**

4.2 Use the Default Account screen to set the default account that you want Mail to use when sending messages.

Switching to another account

When you open the Mail application — in your iPhone 3G's Home screen, tap Mail in the menu bar — you usually see the Inbox folder of your default account. If you have multiple accounts set up on your iPhone 3G and you want to see what's going on with a different account, follow these steps to make the switch:

1. **On the Home screen, tap Mail to open the Mail application.**

2. **Tap the account button that appears in the top-left corner of the screen (but below the status bar).** Keep tapping until the Accounts screen appears, as shown in figure 4.3.

3. **Tap the account you want to work with.** Mail displays a list of the account's folders.

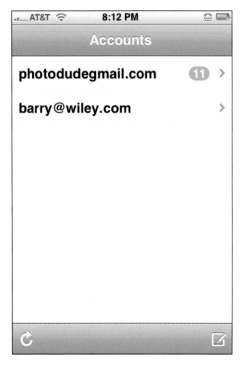

4.3 Use the Accounts screen to choose the e-mail account you want to play with.

Temporarily disabling an account

The Mail application checks for new messages at a regular interval. (We show you how to configure this interval a bit later in this chapter.) If you have several accounts configured in Mail, this incessant checking can put quite a strain on your iPhone 3G battery. To ease up on the juice, you can disable an account temporarily to prevent Mail from checking it for new messages. Here's how:

1. **On the Home screen, tap Settings.** Your iPhone 3G displays the Settings screen.

2. **Tap Mail, Contacts, Calendars to see the Mail settings.**

3. **Tap the account you want to disable.** Your iPhone 3G displays the account's settings.

4. **Tap the Mail switch to turn this setting Off, as shown in figure 4.4.**

5. **Tap Mail to return to the Mail settings screen.**

When you're ready to work with the account again, repeat these steps to turn the Account switch back to On.

4.4 In the account's settings screen, tap the Mail switch to Off.

Deleting an account

If an e-mail account has grown tiresome and boring (or you just don't use it anymore), you should delete it to save storage space, speed up sync times, and save battery power. Follow these steps:

1. **On the Home screen, tap Settings.** The Settings screen appears.

2. **Tap Mail, Contacts, Calendars to get to the Mail settings.**

3. **Tap the account you want to delete.** This opens the account's settings.

4. **At the bottom of the screen, tap Delete Account.** Your iPhone 3G asks you to confirm.

5. **Tap Delete Account.** Your iPhone 3G returns you to the Mail settings screen, and the account no longer graces the Accounts list.

Managing Multiple Devices by Leaving Messages on the Server

In today's increasingly mobile world, it's not unusual to find you need to check the same e-mail account from multiple devices. For example, you might want to check your business account not only using your work computer, but also using your home computer, or using your iPhone 3G while commuting or traveling.

If you need to check e-mail on multiple devices, you can take advantage of how POP e-mail messages are delivered over the Internet. When someone sends you a message, it doesn't come directly to your computer. Instead, it goes to the server that your ISP (or your company) has set up to handle incoming messages. When you ask Apple Mail to check for new messages, it communicates with the POP server to see if any messages are waiting in your account. If so, Mail downloads those messages to your computer, and then instructs the server to delete the copies of the messages that are stored on the server.

The trick, then, is to configure Mail so that it leaves a copy of the messages on the POP server after you download them. That way, the messages are still available when you check messages using another device. Fortunately, the intuitive folks who designed the version of Mail on your iPhone 3G must have understood this, because the program automatically sets up POP accounts to do just that. Specifically, after you download any messages from the POP server to your iPhone 3G, Mail leaves the messages on the server.

Here's a good overall strategy that ensures you can download messages on all your devices, but prevents messages from piling up on the server:

- **Let your main computer be the computer that controls deleting the messages from the server.** In OS X, Mail's default setting is to delete messages from the server after one week, and that's fine.

- **Set up all your other devices — particularly your iPhone 3G — to not delete messages from the server.**

Note

Outlook and Outlook Express always configure POP accounts to delete messages from the server as soon as you retrieve them. You need to fix that. In Outlook, choose Tools ⇨ Account Settings, click the account, click Change, and then click More Settings. Click the Advanced tab and select the Leave a copy of messages on the server check box. In Outlook Express, choose Tools ⇨ Accounts, click your e-mail account, and click Properties. Click the Advanced tab and then select the Leave a copy of messages on server check box.

It's a good idea to check your iPhone 3G POP accounts to ensure they're not deleting messages from the server. To do that, or to use a different setting — such as deleting messages after a week or when you delete them from your Inbox — follow these steps:

1. **On the Home screen, tap Settings.** The Settings screen appears.

2. **Tap Mail, Contacts, Calendars.** Your iPhone 3G opens the Mail, Contacts, Calendars settings screen.

3. **Tap the POP account you want to work with.** The account's settings screen appears.

4. **Near the bottom of the screen, tap Advanced.** Your iPhone 3G displays the Advanced screen.

5. **Tap the Remove from server settings.** The Remove from server screen appears, as shown in figure 4.5.

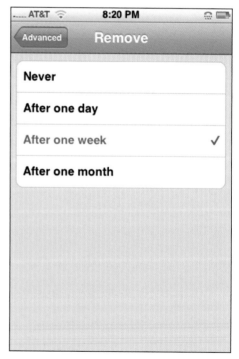

4.5 Use the Remove from server screen to ensure your iPhone 3G is leaving messages on your POP server.

6. **Tap Never.** If you prefer that your iPhone 3G delete messages from the server after seven days, tap one week, instead; if you want your iPhone 3G to delete messages from the server when you delete them from the account's Inbox folder, tap When removed from Inbox, instead.

Fixing Outgoing E-mail Problems by Using a Different Server Port

For security reasons, some Internet service providers (ISPs) insist that all their customers' outgoing mail must be routed through the ISP's Simple Mail Transport Protocol (SMTP) server. This usually is not a big deal if you're using an e-mail account maintained by the ISP, but it can lead to the following problems if you're using an account provided by a third party (such as your Web site host):

- Your ISP might block messages sent using the third-party account because it thinks you're trying to relay the message through the ISP's server (a technique often used by spammers).

- You might incur extra charges if your ISP allows only a certain amount of SMTP bandwidth per month or a certain number of sent messages, whereas the third-party account offers higher limits or no restrictions at all.

- You might have performance problems, with the ISP's server taking much longer to route messages than the third-party host.

You might think that you can solve the problem by specifying the third-party host's SMTP server in the account settings. However, this usually doesn't work because outgoing e-mail is sent by default through port 25; when you use this port, you must also use the ISP's SMTP server.

To work around this problem, many third-party hosts offer access to their SMTP server via a port other than the standard port 25. For example, the .Mac SMTP server (smtp.mac.com) also accepts connections on port 587.

Here's how to configure an e-mail account to use a nonstandard SMTP port:

1. **On the Home screen, tap Settings.** You see the Settings screen.

2. **Tap Mail, Contacts, Calendars.** The Mail, Contacts, Calendars settings screen appears.

3. **Tap the POP account you want to work with.** The account's settings screen appears.

4. **Near the bottom of the screen, tap Advanced.** Your iPhone 3G displays the Advanced screen.

5. **In the Outgoing Mail Server section, tap Server Port.** Your iPhone 3G displays a keypad so you can enter the port number, as shown in figure 4.6.

Configuring Authentication for Outgoing Mail

Because spam is such a big problem these days, many ISPs now require *SMTP authentication* for outgoing mail, which means that you must log on to the SMTP server to confirm that you're the person sending the mail (as opposed to some spammer spoofing your address). If your ISP requires authentication on outgoing messages, you need to configure your e-mail account to provide the proper credentials.

4.6 In the Advanced screen's Outgoing Mail Server area, tap Server Port to enter the new port number to use for outgoing messages.

If you're not too sure about any of this, check with your ISP. If that doesn't work out, by far the most common type of authentication is to specify a username and password (this happens behind the scenes when you send messages). Follow these steps to configure your iPhone 3G e-mail account with this kind of authentication:

1. **On the Home screen, tap Settings.** Your iPhone 3G displays the Settings screen.

2. **Tap Mail, Contacts, Calendars.** The Mail settings screen appears.

3. **Tap the POP account you want to work with.** The account's settings screen appears.

4. **Near the bottom of the screen, tap STMP, and then tap Primary Server.** Your iPhone 3G displays the Advanced screen.

5. **In the Outgoing Mail Server section, tap Authentication.** Your iPhone 3G displays the Authentication screen.

6. **Tap Password.**

7. **Tap the server address to return to the server settings screen.**

8. **In the Outgoing Mail Server section, type your account username in the User Name box, and the account password in the Password box.**

Configuring iPhone 3G to Automatically Check for New Messages

By default, your iPhone 3G only checks for new messages when you tell it to:

1. **On the Home screen, tap Mail to open the Mail application.**

2. **Tap your account button that appears in the top-left corner of the screen (but below the status bar).** Keep tapping until you get to the Accounts screen.

3. **Tap the account you want to work with.** Mail displays a list of the account's folders.

4. **Tap the Inbox folder.** Mail opens the folder and checks for messages.

Genius

While you have an account's Inbox folder open, you can check for messages again by tapping the Refresh icon on the left side of the menu bar.

This is usually the behavior you want, because it limits bandwidth if you're using the cellular network, and it saves battery life. However, if you're busy with something else and you're expecting an important message, you might prefer to have your iPhone 3G check for new messages automatically. Easy money! The Auto-Check feature is happy to handle everything for you. Here's how you set it up:

1. **On the Home screen, tap Settings to display the Settings screen.**

2. **Tap Fetch New Data.** Your iPhone 3G opens the Fetch New Data screen, as shown in figure 4.7.

3. **In the Fetch section, tap the interval you want to use.** For example, if you tap 15 minutes, your iPhone 3G checks all your accounts for new messages every 15 minutes.

When you're ready to return to checking for new messages on your own time, repeat these steps, and when you get to the Fetch New Data screen, tap Manually.

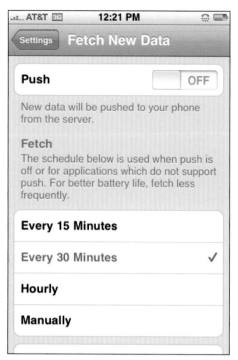

4.7 Use the Fetch New Data screen to configure your iPhone 3G to check for new messages automatically.

Displaying More Messages in Your Inbox Message List

When you open an account's Inbox folder, if you have more than five messages, initially you only see the first five. As you can see in figure 4.8, the reason you see so few messages is that Mail displays for each message the sender, the subject line, and a two-line preview of the message.

Of course, it's not a big whoop to flick through the rest of your messages (it's kind of fun, actually), but if you're looking for (or waiting for) a particular message, it would be nice to see more messages on the screen at once.

Can this be done? Of course, it's the iPhone 3G! The secret is to reduce the number of lines that Mail uses for the message preview. Reduce the preview to a single line, and you now see six messages on the screen; get rid of the preview altogether, and you see a whopping nine messages per screen. (Well, I fibbed: it's actually eight and a half, because the final message shows only the sender's name.)

Follow these steps to reduce the preview size:

1. **On the Home screen, tap Settings.** The Settings screen appears.

2. **Tap Mail, Contacts, Calendars to open the Mail screen.**

3. **Tap Preview.** Your iPhone 3G displays the Preview screen, as shown in figure 4.9.

4. **Tap the number of lines you want to use.** To reduce the preview to a single line, tap 1 Line; to see no preview at all, tap None.

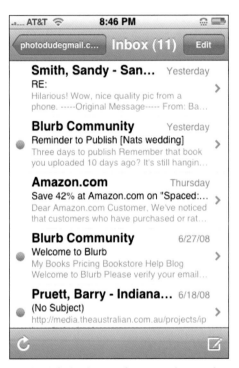

4.8 The default Inbox configuration shows only a maximum of five messages at a time.

4.9 Use the Preview screen to set the number of lines that Mail uses to preview the Inbox messages.

Processing E-mail Faster by Identifying Messages Sent to You

In your iPhone 3G's Mail application, the Inbox folder tells you who sent you each message, but it doesn't tell you to whom the message was sent (that is, which addresses appeared on the To line or the Cc line). No big deal, right? Maybe, maybe not. You see, bulk mailers — we're talking newsletters, mailing lists, and, notoriously, spammers — often don't send messages directly to each person on their subscriber lists. Instead, they use a generic bulk address, which means, significantly, that your e-mail address doesn't appear on the To or Cc lines. That's significant because most newsletters and mailing lists — and *all* spam — are low-priority messages that you can ignore when you're processing a stuffed Inbox.

Okay, great, but what good does all this do you if Mail doesn't show the To and Cc lines? No, it doesn't show those lines, but you *can* configure Mail to show a little icon for messages that were sent directly to you.

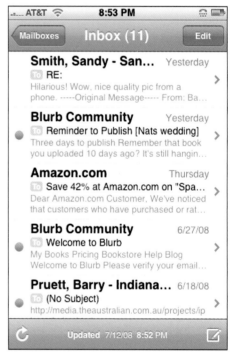

- If the message includes your address in the To field, you see a "To" icon beside the message.

- If the message includes your address in the Cc field, you see a "Cc" icon beside the message.

Neat! Here's how to make this happen:

1. **On the Home screen, tap Settings.** The Settings screen appears.

2. **Tap Mail, Contacts, Calendars.** The Mail, Contacts, Calendars screen appears.

3. **Tap the Show To/Cc Label switch to the On position.**

When you examine your Inbox, you see the To and Cc icons on messages addressed to you, and you don't see either icon on bulk messages, as shown in figure 4.10.

4.10 With the Show To/Cc Label switch turned on, Mail shows you which messages were addressed directly to you.

Placing a Phone Call from an E-mail Message

E-mail messages often include a signature — a line or three of text that appears at the bottom of the message. In business e-mail, a person's signature often includes contact information — what hipster business types like to refer to as their "coordinates" — such as their address (e-mail and postal) and their phone numbers (land and cell). (We show you how to create your own custom iPhone 3G signature a bit later in this chapter.)

In a run-of-the-mill e-mail program, if you wanted to call one of your correspondents based on their signature information, you'd open the message, perhaps jot down the number if there was no phone nearby, and then make the call. As you know, your iPhone 3G isn't run-of-the-mill *anything*, so you can just forget all that rigmarole. Why? Because when it sees a phone number in an e-mail message, your iPhone 3G does something quite smart: it converts that number into a kind of link — the number appears in blue, underlined text, much like a link on a Web page. Figure 4.11 shows an example. Tap the number, tap Call in the dialog box that appears, and your iPhone 3G immediately dials the number for you. Thanks!

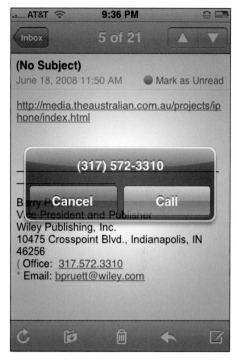

4.11 Your iPhone 3G is savvy enough to convert an e-mail message phone number into a link that you can tap to call the number.

Genius

Your iPhone 3G's Mail application converts phone numbers into fake links that you can tap, so it's not even remotely surprising that Mail also converts Web addresses into actual links. That is, when you tap an address that appears in an e-mail message, your iPhone 3G fires up Safari and takes you to that address. Even better, if the sender includes a link in the message, you can tap and hold the link to see a pop-up bubble that tells you the link address.

E-mailing a Link to a Web Page

The Web is all about finding content that's interesting, educational, and, of course, fun. And if you stumble across a site that meets one or more of these criteria, then the only sensible thing to do is share your good fortune with someone else, right? So, how do you do that? Some sites are kind enough to include an Email This Page link (or something similar), but you can't count on having one of those around. Instead, the usual method is to copy the page address, switch to your e-mail program, paste the address into the message, choose a recipient, and then send the message.

Boy, that sure seems like a ton of work, and in any case it's not something you can do with your iPhone 3G because you can't do the copy-and-paste thing. So are you out of luck? Not a chance (you probably knew that). Your iPhone 3G includes a great little feature that enables you to plop the address of the current Safari page into an e-mail message with just a couple of taps. You then ship out the message and you've made the world a better place.

Here's how it works:

1. **Use Safari to navigate to the site you want to share.**

2. **Tap the + button in the menu bar.** Safari displays a dialog box with several options.

3. **Tap Mail Link to this Page.** This opens a new e-mail message. As you can see in figure 4.12, the new message already includes the page title as the Subject and the page address in the message body.

4. **Choose a recipient for the message.**

5. **Edit the message text as you see fit.**

6. **Tap Send.** Your iPhone 3G fires off the message and returns you to Safari.

4.12 When you tap the Mail Link to this Page option, your iPhone 3G creates a new e-mail message with the page title and address already inserted.

Setting a Minimum Message Font Size

Some people who send e-mails must have terrific eyesight because the font they use for the message text is positively microscopic. Such text is tough to read even on a big screen, but when it's crammed into the iPhone 3G's touchscreen, you'll be reaching for the nearest magnifying glass. Of course, that same touchscreen can also solve this problem: a quick finger spread magnifies the text accordingly.

That's easy enough if you just get the occasional message with nanoscale text, but if a regular correspondent does this, or if your eyesight isn't quite what it used to be (so *all* your messages appear ridiculously teensy), then a more permanent solution might be in order. Your iPhone 3G rides to the rescue once again by letting you configure a minimum font size for your messages. This means that if the message font size is larger than what you specify, your iPhone 3G displays the message as is; however, if the font size is smaller than your specification, your iPhone 3G scales up the text to your minimum size. Your tired eyes will be forever grateful.

Follow these steps to set your minimum font size:

1. **On the Home screen, tap Settings.** The Settings screen appears.

2. **Tap Mail, Contacts, Calendars.** Your iPhone 3G displays the Mail, Contacts, Calendars settings screen.

3. **Tap Minimum Font Size.** The Minimum Font Size screen appears, as shown in figure 4.13.

4. **Tap the minimum font size you want to use: Small, Medium, Large, Extra Large, or Giant.** Mail uses that font size (or larger) when displaying your messages.

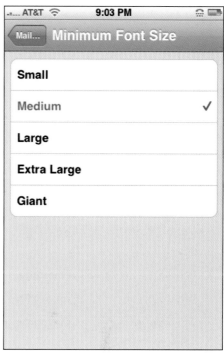

4.13 Use the Minimum Font Size screen to set the smallest text size that you want Mail to use when it displays a message.

Creating a Custom iPhone 3G Signature

E-mail signatures can range from the simple — a signoff such as "Cheers," or "All the best," followed by the sender's name — to baroque masterpieces filled with contact information, snappy quotations, even text-based artwork! On your iPhone 3G, the Mail application takes the simple route by adding the following signature to all your outgoing messages (new messages, replies, and forwards):

```
Sent from my iPhone
```

We really like this signature because it's short, simple, and kinda cool (we, of course, *want* our recipients to know that we're using our iPhone 3G!). If that default signature doesn't rock your world, you can create a custom one that does. Follow these steps:

1. **On the Home screen, tap Settings.** Your iPhone 3G opens the Settings screen.

2. **Tap Mail, Contacts, Calendars.** You see the Mail, Contacts, Calendars settings screen.

3. **Tap Signature.** The Signature screen appears, as shown in figure 4.14.

4. **Type the signature you want to use.**

5. **Tap Mail.** Mail saves your new signature and uses it on all outgoing messages.

4.14 Use the Signature screen to create your custom iPhone 3G e-mail signature.

Caution Mail doesn't give any way to cancel your edits and return to the original signature, so enter your text carefully. If you make a real hash of things, tap Clear to get a fresh start.

Configuring Your Exchange Active Sync Settings

If you have an account on a Microsoft Exchange Server 2003 or 2007 network, and that server has deployed Exchange ActiveSync, then you're all set to have your iPhone 3G and Exchange account synchronized automatically. That's because ActiveSync supports *wireless push* technology, which means that if anything changes on your Exchange server account, that change is immediately synced with your iPhone 3G:

⦿ **E-mail.** If you receive a new message on your Exchange account, ActiveSync immediately displays that message in your iPhone 3G's Mail application.

⦿ **Contacts.** If someone at work adds or edits data in the server address book, those changes are immediately synced to your iPhone 3G Contacts list.

⦿ **Calendar.** If someone at work adds or edits an appointment in your calendar, or if someone requests a meeting with you, that data is immediately synced with your iPhone 3G's Calendar application.

ActiveSync works both ways, too, so if you send e-mail messages, add contacts or appointments, or accept meeting requests, your server account is immediately updated with the changes. And all this data whizzing back and forth is safe, because it's sent over a secure connection.

Your iPhone 3G also gives you a few options for controlling ActiveSync, and the following steps show you how to set them:

1. **On the Home screen, tap Settings.** Your iPhone 3G opens the Settings screen.

2. **Tap Mail, Contacts, Calendars to open the Mail, Contacts, Calendars settings.**

3. **Tap your Exchange account.** The Exchange ActiveSync screen appears, as shown in figure 4.15.

4. **To sync your Exchange e-mail account, tap the Mail On/Off switch to the On position.**

5. **To sync your Exchange address book, tap the Contacts On/Off switch to the On position, and then click Sync.**

6. **To sync your Exchange calendar, tap the Calendars On/Off switch to the On position, and then click Sync.**

7. **To control the amount of time that gets synced on your e-mail account, tap Mail Days to Sync and then tap the number of days, weeks, or months you want to sync.**

4.15 Use the Exchange ActiveSync screen to customize your iPhone 3G's ActiveSync support.

93

How Can I Take Control of Syncing My iPhone 3G?

Your iPhone 3G can function perfectly well on its own. After all, you can use it to create your own bookmarks, e-mail accounts, contacts, and appointments; you can download music and other media from the iTunes store; and you can take your own photos using the iPhone 3G's built-in camera. If you want to use your iPhone 3G as a stand-alone device, no one can stop you, not even Steve Jobs himself (we think). Yes, you can do that, but we're not sure *why* you'd want to. With all of the iPhone 3G's great synchronization features, tons of that useful and fun content on your computer can also be shared with your iPhone 3G. This chapter shows you how to master syncing your iPhone 3G and your Mac or Windows computer.

Connecting Your iPhone 3G to Your Computer

In an ideal world, your iPhone 3G and your computer would be able to converse back and forth using some sort of wireless connection. That day may come, but for now we're stuck in a world of wires, which means you need to make a physical connection between your iPhone 3G and your computer to do the syncing thing.

You've got a couple of ways to make the connection:

- **USB cable.** Take the USB cable that comes with your iPhone 3G and attach the USB connector to a free USB port on your Mac or Windows PC, and then attach the dock connector to the iPhone 3G.

- **Dock.** If you shelled out the bucks for the optional iPhone 3G dock, first plug it in to a power outlet. Using your iPhone 3G's USB cable, attach the USB connector to a free USB port on your Mac or Windows PC, and attach the dock connector to the dock. Now insert your iPhone 3G into the dock's cradle.

Syncing Your iPhone 3G Automatically

Start with the look-ma-no-hands syncing scenario where you don't have to pay any attention in the least: automatic syncing. If the amount of iPhone 3G-friendly digital content you have on your Mac or Windows PC is less than the capacity of your iPhone 3G, then you have no worries because you know it's all going to fit. So all you have to do is turn on your iPhone 3G and then connect it to your computer:

Note If a call comes in while you're syncing, your iPhone 3G cancels the sync automatically so that you can take the call.

That's it! iTunes opens automatically and begin syncing your iPhone 3G (and, as an added bonus, also begins charging your iPhone 3G's battery). Your iPhone 3G displays the Sync in Progress screen while the sync runs, and then returns you to the Home screen when the sync is complete. Note that you can't use your iPhone 3G while the sync is running.

However, one of the iPhone 3G's nicest features is its willingness to be rudely interrupted in midsync. Figure 5.1 shows the Sync in Progress screen. Check out the Slide to cancel slider at the bottom. If you ever need to bail out of the sync to, say, make a call, drag the slider to the right. iTunes dutifully cancels the sync so you can go about your business. To restart the sync, click the Sync button in iTunes.

5.1 If you need to use your iPhone 3G in midsync, drag the Slide to cancel slider to the right to stop the sync.

Bypassing the automatic sync

What do you do if you want to connect your iPhone 3G to your computer, but you don't want it to sync? We're not talking here about switching to manual syncing full time (we'll get to that in a second). Instead, we're talking about bypassing the sync one time only. For example, you might want to connect your iPhone 3G to your computer just to charge it (assuming you either don't have the optional dock or you don't have it with you). Or perhaps you just want to use iTunes to eyeball how much free space is left on your iPhone 3G or to check for updates to the iPhone 3G software.

Genius

We've found that syncing can sometimes fail if your iPhone 3G is open to an application's settings screen when you launch the sync. Make sure no settings are open before trying to sync.

Whatever the reason, you can tell iTunes to hold off the syncing this time only by using one of the following techniques:

- **Mac.** Connect the iPhone 3G to the Mac and then quickly press and hold the Option and ⌘ keys.

- **Windows.** Connect the iPhone 3G to the Windows PC and then quickly press and hold the Ctrl and Shift keys.

When you see that iTunes has added your iPhone 3G to the Devices list, you can release the keys.

Troubleshooting automatic syncing

Okay, so you connect your iPhone 3G to your computer and then nothing. iTunes doesn't wake from its digital slumbers or, if iTunes is already running, it sees the iPhone 3G but refuses to start syncing. What's up with that?

It could be a couple of things. First, connect your iPhone 3G, switch to iTunes on your computer, and then click your iPhone 3G in the Devices list. On the Summary tab (see figure 5.2), make sure the Automatically sync when this iPhone is connected check box is selected.

If that check box was already selected, then you need to delve a bit deeper to solve the mystery. Follow these steps:

1. **Open the iTunes preferences:**
   - **Mac.** Choose iTunes ➪ Preferences, or press ⌘+, (period).
   - **Windows.** Choose Edit ➪ Preferences, or press Ctrl+, (period).

2. **Click the Syncing tab.**

3. **Deselect the Disable automatic syncing for all iPhones and iPods check box.**

4. **Click OK to put the new setting into effect and enable automatic syncing one again.**

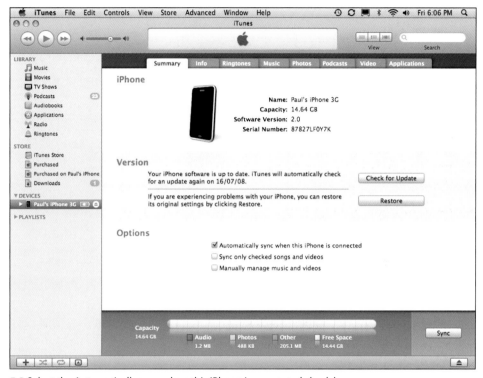

5.2 Select the Automatically sync when this iPhone is connected check box.

Giving Your iPhone 3G a Snappy Name

This isn't necessarily a syncing topic, but we thought that while you're in iTunes, you might want to give your iPhone 3G a proper name, one that's a tad more interesting than the boring "iPhone 3G" that passes for the default name. Here's what you do:

1. **Double-click your iPhone 3G in the Devices list.** iTunes forms a text box around the name.

2. **Type the name you want to use.** You can use any characters you want, and the name can be as long as you want (although you might want to use no more than about 15 or 16 characters to ensure the name doesn't get cut off in the Devices list).

3. **Press Return or Enter to save the new name.**

As soon as you press Return or Enter, iTunes connects to your iPhone 3G and saves the name on the phone. This way, even if you connect your iPhone 3G to another computer, that machine's version of iTunes will show your custom iPhone 3G name.

Syncing Your iPhone 3G Manually

One fine day, you'll be minding your own business and performing what you believe to be a routine sync operation when a dialog box like the one shown in figure 5.3 rears its nasty head.

Groan! This most unwelcome dialog box means just what it says: There's not enough free space on your iPhone 3G to sync all the content from your computer. You've got a couple of ways to handle this:

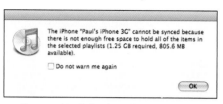

The iPhone "Paul's iPhone 3G" cannot be synced because there is not enough free space to hold all of the items in the selected playlists (1.25 GB required, 805.6 MB available).

☐ Do not warn me again

OK

5.3 You see this dialog box if iTunes can't fit all of your stuff on your iPhone 3G.

- **Remove some of the content from your computer.** This is a good way to go if your iPhone 3G is really close to having enough space. For example, the dialog box says your computer wants to send 100MB of data, but your iPhone 3G has only 98MB of free space. Get rid of a few megabytes of stuff on your computer, and you're back in the sync business.

- **Synchronize your iPhone 3G manually.** This means that you no longer sync everything on your computer. Instead, you handpick which playlists, podcasts, audiobooks, and so on are sent to your iPhone 3G. It's a bit more work, but it's the way to go if there's a big difference between the amount of content on your computer and the amount of space left on your iPhone 3G.

The rest of this chapter shows you how to manually sync the various content types: contacts, calendars, e-mail, bookmarks, music, podcasts, audiobooks, photos, and videos.

Syncing Information with Your iPhone 3G

If you step back a pace or two to take in the big picture, you see that your iPhone 3G deals with two broad types of data: media — all that audio and video stuff — and information — contacts, appointments, e-mail, and Web sites. You need both types of data to get the most out of your iPhone 3G investment, and happily, both types of data are eminently syncable. We'll get to the media syncing portion of the show a bit later. For now, the next few sections show you how to take control of syncing your information between your iPhone 3G and your computer.

Syncing your contacts

Although you can certainly add contacts directly on your iPhone 3G — and we show you how to do just that in Chapter 7 — adding, editing, grouping, and deleting contacts is a lot easier on a computer. So a good way to approach contacts is to manage them on your Mac or Windows PC, and then sync your contacts with your iPhone 3G.

However, do you really need to sync *all* your contacts? For example, if you only use your iPhone 3G to contact friends and family, then why clog your phone's Contacts list with work contacts? We don't know!

You can control which contacts are sent to your iPhone 3G by creating groups of contacts, and then syncing only the groups you want. Here are some quickie instructions for creating groups:

- **Address Book (Mac).** Choose File ⇨ New Group, type the group name, and then press Return. Now populate the new group by dragging and dropping contacts on it.

- **Address Book (Windows XP).** Choose File ⇨ New Group, type the group name, and then click Select Members. For each person you want in the group, choose the contact and click Select. When you're done, click OK, and then click OK again.

- **Contacts (Windows Vista).** Click New Contact Group, type the group name, and then click Add to Contact Group. Choose all the contacts you want in the group and then click Add. Click OK.

Note

If you're an Outlook user, note that iTunes doesn't support Outlook-based contact groups, so you're stuck with syncing everyone in your Outlook Contacts folder.

With your group (or groups) all figured out, follow these steps to sync your contacts with your iPhone 3G:

1. **Connect your iPhone 3G to your computer.**

2. **In iTunes, click your iPhone 3G in the Devices list.**

3. **Click the Info tab.**

4. **Select the Sync Address Book contacts check box.**

5. **Select an option:**

   - **All contacts.** Select this option to sync all your Address Book contacts.

- **Selected groups.** Select this option to sync only the groups you pick. In the group list, select the check box beside each group that you want to sync, as shown in figure 5.4.

6. **If you want to make the sync a two-way street, select the Put new contacts created on this iPhone into the group option, and then choose a group from the menu.**

7. **If you have a Yahoo! account and you also want your Yahoo! Address Book contacts in on the sync, use one of the following techniques:**

   - **Mac.** Select the Sync Yahoo! Address Book contacts check box, click Agree, type your Yahoo! ID and password, and click OK.

   - **Windows.** Use the Sync contacts with list to choose Yahoo! Address Book, click Agree, type your Yahoo! ID and password, and click OK

5.4 You can sync selected Address Book groups with your iPhone 3G..

8. **If you have a Google account and you also want your Google Contacts in on the sync, use one of the following techniques:**

   - **Mac.** Select the Sync Google Contacts check box, click Agree, type your Google and password, and click OK.

   - **Windows.** Use the Sync contacts with list to choose Google Contacts, click Agree, type your Google ID and password, and click OK.

9. **Click Apply.** iTunes syncs the iPhone 3G using your new contacts settings.

Syncing your calendar

When you're tripping around town with your trusty iPhone 3G at your side, you certainly don't want to be late if you've got a date. The best way to ensure that you don't miss an appointment, meeting, or rendezvous is to always have the event details at hand, which means adding those details to your iPhone 3G's Calendar. You could add the appointment to Calendar right on the iPhone 3G (a technique we take you through in Chapter 7), but it's easier to create it on your computer and then sync it to your iPhone 3G. This gives you the added advantage of having the appointment listed in two places, so you're sure to arrive on time.

Most people sync all the appointments, but it's not unusual to keep track of separate schedules — for example, business and personal. You can control which schedule is synced to your iPhone 3G by creating separate calendars and then syncing only the calendars you want. In your Mac's iCal application, choose File ⇨ New Calendar, type the calendar name, and then press Return.

Note   Although you can create extra calendars in Outlook, iTunes doesn't recognize them, so you have to sync everything in your Outlook Calendar folder. Also, iTunes doesn't support Windows Calendar (available with Windows Vista), so you're out of luck if you use that to manage your schedule.

Now follow these steps to sync your calendar with your iPhone 3G:

1. **Connect your iPhone 3G to your computer.**

2. **In iTunes, click your iPhone 3G in the Devices list.**

3. **Click the Info tab.**

4. **Turn on calendar syncing by using one of the following techniques:**

   - **Mac.** Select the Sync iCal calendars check box.

   - **Windows.** Select the Sync calendars with check box, and then use the list to choose the program you want to use (such as Outlook).

5. **Select an option:**

   - **All calendars.** Select this option to sync all your calendars.

   - **Selected calendars.** Select this option to sync only the calendars you pick. In the calendar list, select the check box beside each calendar that you want to sync, as shown in figure 5.5.

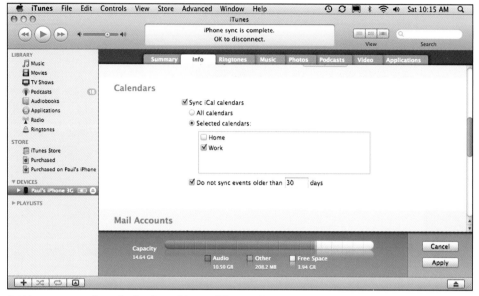

5.5 You can sync selected calendars with your iPhone 3G.

6. **To control how far back the calendar sync goes, select the Do not sync events older than *X* days check box, and then type the number of days of calendar history you want to see on your iPhone 3G.**

7. **Click Apply.** iTunes syncs the iPhone 3G using your new calendar settings.

Syncing your e-mail account

By far the easiest way to configure your iPhone 3G with an e-mail account is to let iTunes do all the heavy lifting for you. That is, if you've got an existing account already up and running — whether it's a Mail account on your Mac, or an Outlook or Outlook Express account on your Windows PC — you can convince iTunes to gather all the account details and pass them along to your iPhone 3G. Here's how it works:

1. **Connect your iPhone 3G to your computer.**

2. **In the iTunes sources list, click the iPhone 3G.**

3. **Click the Info tab.**

4. **In the Mail Accounts section, use one of the following techniques:**

   ● **Mac.** Select the Sync selected Mail accounts check box, and then select the check box beside each account you want to add to iPhone 3G, as shown in figure 5.6.

● **Windows.** Select the Sync selected mail accounts with check box, select your e-mail pro-gram from the drop-down list, and then select the check box beside each account you want to add to iPhone 3G.

5.6 Make sure you select the Sync selected Mail account check box and at least one account in the list.

5. **Click Apply.** You may see a message asking if AppleMobileSync can be allowed access to your keychain (your Mac's master password list).

6. **If you see that message, click Allow.** iTunes begins syncing the selected e-mail account settings from your computer to your iPhone 3G.

Syncing your bookmarks

The easiest way to get bookmarks for your favorite sites in your iPhone 3G is to take advantage of your best bookmark resource: the Safari browser on your Mac (or Windows), or the Internet Explorer browser on your Windows PC (which calls them favorites). You've probably used those browsers for a while and have all kinds of useful and fun bookmarked sites at your metaphorical fingertips. To get those bookmarks at your literal fingertips — that is, on your iPhone 3G — you need to include book-marks as part of the synchronization process between the iPhone 3G and iTunes.

Bookmark syncing is turned on by default, but you should follow these steps to make sure:

1. **Connect your iPhone 3G to your computer.**

2. **In the iTunes sources list, click the iPhone 3G.**

Caution

Having used Safari or Internet Explorer for a while means having lots of great sites bookmarked, but it likely also means that you've also got lots of digital dreck — sites you no longer visit, or that have gone belly-up. Before synchronizing your bookmarks with the iPhone 3G, consider taking some time to clean up your existing bookmarks. You'll thank yourself in the end.

Genius

What's that? You've already synced your bookmarks to your iPhone 3G and you now have a bunch of useless sites clogging up Mobile Safari's bookmark arteries? Not a problem! Return to your desktop Safari (or Internet Explorer), purge the bogus bookmarks, and then sync your iPhone 3G using the following steps. Any bookmarks you blew away also get trashed from iPhone 3G.

3. **Click the Info tab.**

4. **Scroll down to the Web Browser section, and then use one of the following techniques:**

   ● **Mac.** Select the Sync Safari bookmarks check box, as shown in figure 5.7.

   ● **Windows.** Select the Sync bookmarks with check box, and then select your Web browser from the drop-down list.

5. **Click Apply.** iTunes begins syncing the bookmarks from your computer to your iPhone 3G.

5.7 Make sure the Sync Safari bookmarks check box is selected.

Merging data from two or more computers

Long gone are the days when our information resided on a single computer. Now it's common to have a desktop computer (or two) at home, a work computer, and perhaps a notebook to take on the road. It's nice to have all that digital firepower, but it creates a big problem: You end up with contacts, calendars, and other information scattered over several machines. How are you supposed to keep track of it all?

Apple's latest solution is MobileMe, which promises seamless information integration across multiple computers (Mac and Windows) and, of course, the iPhone 3G, and that's the topic we cover in Chapter 9.

If you don't have a MobileMe account, you can still achieve a bit of data harmony. That's because iTunes offers the welcome ability to *merge* information from two or more computers on the iPhone 3G. For example, if you have contacts on your home computer, you can sync them with your iPhone 3G. If you have a separate collection of contacts on your notebook, you can also sync them with your iPhone 3G, but iTunes gives you two choices:

- **Merge Info.** With this option, your iPhone 3G keeps the information synced from the first computer and merges it with the information synced from the second computer.

- **Replace Info.** With this option, your iPhone 3G deletes the information synced from the first computer and replaces it with the information synced from the second computer.

Here are the general steps to follow to set up your merged information:

1. **Sync your iPhone 3G with information from one computer.** This technique works with contacts, calendars, e-mail accounts, and bookmarks.

2. **Connect your iPhone 3G to the second computer.**

3. **In iTunes, click your iPhone 3G in the Devices list.**

4. **Click the Info tab.**

5. **Select the Sync check boxes that correspond to information already synced on the first computer.** For example, if you synced contacts on the first computer, select the Sync contacts check box.

6. **Click Apply.** iTunes displays a dialog box like the one shown in figure 5.8.

5.8 You can merge contacts, calendars, e-mail accounts, and bookmarks from two or more computers.

7. **Click Merge Info.** iTunes syncs your iPhone 3G and merges the computer's information with the existing information from the first computer.

Handling sync conflicts

When you sync information between your iPhone 3G and a computer, any edits you make to that information are included in the sync. For example, if you change someone's e-mail address on your iPhone 3G, the next time you sync, iTunes updates the e-mail address on the computer, which is exactly what you want.

However, what if you already changed that person's address on the computer? If you made the same edit, then it's no biggie because there's nothing to sync. But what if you made a *different* edit? Ah, that's a problem, because now iTunes doesn't know which version has the correct information. In that case, it shrugs its digital shoulders and passes off the problem to a program called Conflict Resolver, which displays the dialog box shown in figure 5.9.

5.9 If you make different edits to the same bit of information on your iPhone 3G and your computer, the Conflict Resolver springs into action.

If you want to deal with the problem now, click Review Now. Conflict Resolver then offers you the details of the conflict. For example, in figure 5.10 you can see that a contact's work e-mail address is different in Address Book and on the iPhone 3G. To settle the issue once and for all (you hope), click the correct version of the information, and then click Done. When Conflict Resolver tells you it will fix the problem during the next sync, click Sync Now to make it happen right away.

5.10 Review Now shows you the details of any conflilcts.

Handling large iPhone 3G-to-computer sync changes

Syncing works both ways, meaning that not only does your iPhone 3G receive content from your computer, but your computer also receives content from your iPhone 3G. For example, if you create any bookmarks, contacts, or appointments on your iPhone 3G, those items get sent to your computer during the sync.

However, it's implied that the bulk of the content flows from your computer to your iPhone 3G, which makes sense because for most things it's easier to add, edit, and delete stuff on the computer. So that's why if you make lots of changes to your iPhone 3G content, iTunes displays a warning that the sync is going to make lots of changes to your computer content. The threshold is five percent, which means that if the sync will change more than five percent of a particular type of content on your computer — such as bookmarks or calendars — the warning appears. For example, figure 5.11 shows the Sync Alert dialog box you see if the sync will change more than five percent of your computer's bookmarks.

If you're expecting this (because you *did* change lots of stuff on your iPhone 3G), click the Sync *Whatever* button (where *Whatever* is the type of data: Bookmarks, Calendars, and so on). If you're not sure, click Show Details to see what the changes are. If you're still scratching your head, click Cancel to skip that part of the sync.

5.11 iTunes warns you if the sync will mess with more than five percent of your computer's content.

If you're running iTunes for Windows, you can either turn off this warning or adjust the threshold. (For some unfathomable reason, iTunes for the Mac doesn't offer this handy option.) Follow these steps:

1. **Choose Edit ⇨ Preferences, or press Ctrl+,.** The iTunes dialog box comes aboard.

2. **Click the Syncing tab.**

3. **If you want to disable the sync alerts altogether, deselect the Warn when check box.** Otherwise, leave that check box selected and move on to step 4.

4. **Use the Warn when *percent* of the data on the computer will be changed list to set the alert threshold, where *percent* is one of the following:**

   - **any.** Select this option to see the sync alert whenever syncing with the iPhone 3G will change data on your computer. iPhone 3G syncs routinely modify data on the computer, so be prepared to see the alerts every time you sync. (Of course, that may be exactly what you want.)

   - **more than 5%.** This is the default setting where you only see the alert if more than five percent of some data type will be changed on the computer during the sync.

- **more than 25%.** Select this option to see the alert only when the sync will change more than 25 percent of some data type on the computer.

- **more than 50%.** Select this option to see the alert only when the sync will change more than 50 percent of some data type on the computer.

5. **Click OK to put the new settings into effect.**

Replacing your iPhone 3G's data with fresh info

After you know what you're doing, syncing contacts, calendars, e-mail accounts, and bookmarks to your iPhone 3G is a relatively bulletproof procedure that should happen without a hitch each time. Of course, this is technology we're dealing with here, so hitches do happen every now and then, and as a result you might end up with corrupt or repeated information on your iPhone 3G.

Or perhaps you've been syncing your iPhone 3G with a couple of different computers, and you decide to cut one of the computers out of the loop and revert to just a single machine for all your syncs.

In both these scenarios, what you need to do is replace the existing information on your iPhone 3G with a freshly baked batch of data. Fortunately, iTunes has a feature that lets you do exactly that. Here's how it works:

1. **Connect your iPhone 3G to your computer.**

2. **In the iTunes sources list, click the iPhone 3G.**

3. **Click the Info tab.**

4. **Select the Sync check boxes for each type of information you want to work with (contacts, calendars, e-mail accounts, or bookmarks).** If you don't select a check box, iTunes won't replace that information on your iPhone 3G. For example, if you like your iPhone 3G bookmarks just the way they are, don't select the Sync bookmarks check box.

5. **In the Advanced section, select the check box beside each type of information you want to replace.** As shown in figure 5.12, there are four check boxes: Contacts, Calendars, Mail Account, and Bookmarks.

6. **Click Apply.** iTunes replaces the selected information on your iPhone 3G.

Note   If a check box in the Advanced section is disabled, it's because you didn't select the corresponding Sync check box. For example, in figure 5.11 you see that the Sync selected Mail accounts check box is deselected, so in the Advanced section the Mail Accounts check box is disabled.

5.12 Use the check boxes in the Advanced section to decide which information you want replaced on your iPhone 3G.

Syncing Media with Your iPhone 3G

The brainy Phone application and the sleek Safari browser may get the lion's share of kudos for the iPhone 3G, but many people reserve their rave reviews for its iPod player. The darn thing is just so versatile: It can play music, of course, but it also happily cranks out audiobooks and podcasts on the audio side, and music videos, movies, and TV shows on the video side. Ear candy and eye candy in one package!

If there's a problem with this digital largesse, it's that the iPod player might be *too* versatile. Even if you have a big 16GB iPhone 3G, you may still find its confines a bit cramped, particularly if you're also loading up your iPhone 3G with photos, contacts, and calendars, and you just can't seem to keep your hands out of the App Store cookie jar.

All this means that you probably have to pay a bit more attention when it comes to syncing your iPhone 3G, and the following sections show you how to do just that.

Syncing music and music videos

The iPhone 3G iPod is a digital music player at heart, so you've probably loaded up your iPhone 3G with lots of audio content and lots of music videos. To get the most out of your iPod's music and video capabilities, you need to know all the different ways you can synchronize these items. For example, if you use your iPhone 3G's iPod primarily as a music player and the iPhone 3G has more disk capacity than you need for all of your digital audio, feel free to throw all your music onto the player. On the other hand, your iPhone 3G might not have much free space, or you might only want certain songs and videos on the player to make it easier to navigate. Not a problem! You need to configure iTunes to sync only the songs that you select.

Before getting to the specific sync steps, you need to know that there are three ways to manually sync music and music videos:

● **Playlists.** With this method, you specify the playlists that you want iTunes to sync. Those playlists also appear on your iPhone 3G's iPod player. This is by far the easiest way to manually sync music and music videos, because you usually just have a few playlists to select. The downside is that if you have large playlists and you run out of space on your iPhone 3G, the only way to fix the problem is to remove an entire playlist. Another bummer: with this method, you can only sync *all* of your music videos, or *none* of your music videos.

Genius Something we like about syncing playlists is that you can estimate in advance how much space your selected playlists will usurp on the iPhone 3G. In iTunes, click the playlist and then examine the status bar, which tells you the number of songs in the playlist, the total duration, and, most significantly for our purposes, the total size of the playlist.

● **Check boxes.** With this method, you specify which songs and music videos get synced by selecting the little check boxes that appear beside every song and video in iTunes. This is fine-grained syncing for sure, but because your iPhone 3G can hold thousands of songs, it's also a lot of work.

● **Drag-and-drop.** With this method, you click and drag individual songs and music videos, and drop them on your iPhone 3G's icon in the iTunes Devices list. This is an easy way to get a bunch of tracks on your iPhone 3G quickly, but iTunes doesn't give you any way of tracking which tracks you've dragged and dropped.

Here are the steps to follow to sync music and music videos using playlists:

1. **In iTunes, click your iPhone 3G in the Devices list.**

2. **Click the Music tab.**

3. **Select the Sync music check box.** iTunes asks you to confirm that you want to sync music.

4. **Click Sync Music.**

5. **Select the Selected playlists option.**

6. **Select the check box beside each playlist you want to sync, as shown in figure 5.13.**

7. **Select the Include music videos check box if you also want to add your music videos into the sync mix.**

8. **Click Apply.** iTunes syncs your iPhone 3G using the new settings.

5.13 Select the Selected playlists option and then select the playlists you want to sync.

Here are the steps to follow to sync using the check boxes that appear beside each track in your iTunes Music library:

1. **Click your iPhone 3G in the Devices list.**

2. **Click the Summary tab.**

3. **Select the Sync only checked songs and videos check box.**

4. **Click Apply.** If iTunes starts syncing your iPhone 3G, drag the Slide to cancel slider on the iPhone 3G to stop it.

5. **Either click Music in the Library list or click a playlist that contains the tracks you want to sync.** If a track's check box is selected, iTunes syncs the track with your iPod. If a track's check box is deselected, iTunes doesn't sync the track with your iPod; if the track is already on your iPod, iTunes removes the track.

Genius

What do you do if you only want to select a few tracks from a large playlist? Waste a big chunk of your life deselecting a few hundred check boxes? Pass. Here's a better way: Press ⌘+A (Mac) or Ctrl+A (Windows) to select every track, right-click (or Ctrl+click on your Mac) any track, and then click Uncheck Selection. Voila! iTunes deselects every track in seconds flat. Now you can select just the tracks you want. You're welcome.

6. **In the Devices list, click your iPhone 3G.**

7. **Click the Summary tab.**

8. **Click Sync.** iTunes syncs just the checked tracks.

You can also configure iTunes to let you drag tracks from the Music library (or any playlist) and drop them on your iPhone 3G. Here's how this works:

1. **Click your iPhone 3G in the Devices list.**

2. **Click the Summary tab.**

3. **Select the Manually manage music and videos check box.**

4. **Click Apply.** If iTunes starts syncing your iPhone 3G, drag the Slide to cancel slider on the iPhone 3G to stop it.

5. **Either click Music in the Library list or click a playlist that contains the tracks you want to sync.**

6. **Choose the tracks you want to sync:**

   ● If all the tracks are together, Shift+Click the first track, hold down Shift, and then click the last track.

   ● If the tracks are scattered all over the place, hold down ⌘ (Mac) or Ctrl (Windows) and click each track.

7. **Click and drag the selected tracks to the Devices list and drop them on the iPhone 3G icon.** iTunes syncs the selected tracks.

Note

When you select the Manually manage music and videos check box, iTunes automatically deselects the Sync music check box in the Music tab. However, iTunes doesn't mess with the music on your iPhone 3G. Even when it syncs after a drag and drop, it only adds the new tracks, it doesn't delete any of your phone's existing music.

Caution

If you decide to return to playlist syncing by selecting the Sync music check box in the Music tab, iTunes removes all tracks that you added to your iPhone 3G via the drag-and-drop method.

Syncing podcasts

In many ways, podcasts are the most problematic of the various media you can sync with your iPod. Not that the podcasts themselves pose any concern. Quite the contrary: They're so addictive that it's not unusual to collect them by the dozens. Why is that a problem? Because most professional podcasts are at least a few megabytes in size, and many are tens of megabytes. A large enough collection can put a serious dent in your iPhone 3G's remaining storage space.

All the more reason to take control of the podcast syncing process. Here's how you do it:

1. **Click your iPhone 3G in the Devices list.**

2. **Click the Podcasts tab.**

3. **Select the Sync check box and choose an option from the drop-down menu:**

   - **All.** Choose this item to sync every podcast.

   - **1 Most Recent.** Choose this item to sync the most recent podcast.

   - **3 Most Recent.** Choose this item to sync the three most recent podcasts.

   - **5 Most Recent.** Choose this item to sync the five most recent podcasts.

   - **10 Most Recent.** Choose this item to sync the ten most recent podcasts.

   - **All Unplayed.** Choose this item to sync all the podcasts you haven't yet played.

   - **1 Most Recent Unplayed.** Choose this item to sync the most recent podcast that you haven't yet played.

   - **3 Most Recent Unplayed.** Choose this item to sync the three most recent podcasts that you haven't yet played.

   - **5 Most Recent Unplayed.** Choose this item to sync the five most recent podcasts that you haven't yet played.

   - **10 Most Recent Unplayed.** Choose this item to sync the ten most recent podcasts that you haven't yet played.

   - **1 Least Recent Unplayed.** Choose this item to sync the oldest podcast that you haven't yet played.

- **3 Least Recent Unplayed.** Choose this item to sync the three oldest podcasts that you haven't yet played.

- **5 Least Recent Unplayed.** Choose this item to sync the five oldest podcasts that you haven't yet played.

- **10 Least Recent Unplayed.** Choose this item to sync the ten oldest podcasts that you haven't yet played.

- **All New.** Choose this item to sync all the podcasts published since the last sync.

- **1 Most Recent New.** Choose this item to sync the most recent podcast published since the last sync.

- **3 Most Recent New.** Choose this item to sync the three most recent podcasts published since the last sync.

- **5 Most Recent New.** Choose this item to sync the five most recent podcasts published since the last sync.

- **10 Most Recent New.** Choose this item to sync the ten most recent podcasts published since the last sync.

- **1 Least Recent New.** Choose this item to sync the oldest podcast published since the last sync.

- **3 Least Recent New.** Choose this item to sync the three oldest podcasts published since the last sync.

- **5 Least Recent New.** Choose this item to sync the five oldest podcasts published since the last sync.

- **10 Least Recent New.** Choose this item to sync the ten oldest podcasts published since the last sync.

Note

A podcast episode is unplayed if you haven't yet played at least part of the episode either in iTunes or your iPhone 3G. If you play an episode on your iPhone 3G, the player sends this information to iTunes when you next sync. Even better, your iPhone 3G also lets iTunes know if you paused in the middle of an episode; when you play that episode in iTunes, it starts at the point where you left off.

Genius

To mark a podcast episode as unplayed, in iTunes choose the Podcasts library, right-click (or Ctrl+Click on your Mac) the episode, and then choose Mark as New.

4. **Select one of these options:**

   - **All podcasts.** Select this option to sync all your podcasts with your iPod.

   - **Selected podcasts.** Select this option to sync specific podcasts with your iPod. Select the check boxes for the items you want to sync, as shown in figure 5.14.

5. **Click Apply.** iTunes syncs the iPhone 3G using your new podcast settings.

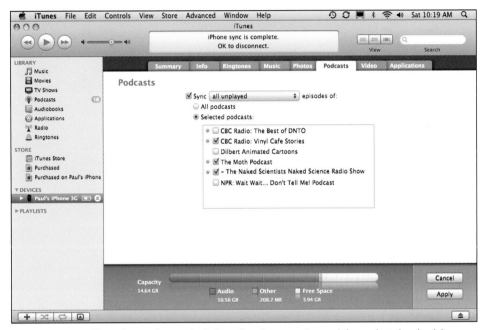

5.14 To sync specific podcasts, choose the Selected podcasts option and then select the check boxes for each podcast you want synced.

Syncing audiobooks

The iTunes sync settings for your iPhone 3G have tabs for Music, Photos, Podcasts, and Video, but not one for Audiobooks. What's up with that? It's not, as you might think, some sort of anti-book conspiracy, or even forgetfulness on Apple's part. Instead, iTunes treats audiobook content as a special type of playlist, which, confusingly, doesn't appear in the iTunes Playlists section. To get audiobooks on your iPhone 3G, follow these steps:

1. **Click your iPhone 3G in the Devices list.**
2. **Click the Music tab.**

3. **Select the Sync music check box, if you haven't done so already.** iTunes asks you to confirm that you want to sync music.

4. **Click Sync Music.**

5. **Select the Selected playlists option.**

6. **Select the check box beside Audiobooks.**

7. **Click Apply.** iTunes syncs your audiobooks to your iPhone 3G.

If you've opted to manually manage your music and video, then you need to choose the Audiobooks category of the iTunes Library, and then drag and drop on your iPhone 3G the audiobooks you want to sync.

Syncing movies

It wasn't all that long ago when technology prognosticators and pundits laughed at the idea of people watching movies on a 2-inch by 3-inch screen. Who could stand to watch even a music video on such a tiny screen? The pundits were wrong, of course, because now it's not at all unusual for people to use their iPhone 3Gs to watch not only music videos, but also short films, animated shorts, and even full-length movies.

The major problem with movies is that their file size tends to be quite large — even short films lasting just a few minutes weigh in at dozens of megabytes, and full-length movies are several gigabytes. Clearly there's a compelling need to manage your movies to avoid filling up your iPhone 3G and leaving no room for the latest album from your favorite band.

Follow these steps to configure and run the movie synchronization:

1. **Click your iPhone 3G in the Devices list.**

2. **Click the Video tab.**

3. **Select the Sync movies check box.**

4. **Select the check box beside each movie you want to sync.**

5. **Click Apply.** iTunes syncs the iPhone 3G using your new movie settings.

Genius

If you download a music video from the Web and then import it into iTunes (by choosing File ▷ Import), iTunes adds the video to its Movies library. To display it in the Music library instead, open the Movies library, right-click (or Ctrl+Click on the Mac) the music video, and then click Get Info. Click the Video tab and then use the Kind list to choose Music Video. Click OK. iTunes moves the music video to the Music folder.

Syncing TV show episodes

If the average iPhone 3G is at some risk of being filled up by a few large movie files, it probably is at grave risk of being overwhelmed by a large number of TV show episodes. A single half-hour episode will eat up approximately 250MB, so even a modest collection of shows will consume multiple gigabytes of precious iPhone 3G disk space.

This means it's crucial to monitor your collection of TV show episodes and keep your iPhone 3G synced with only the episodes you need. Fortunately, iTunes gives you a decent set of tools to handle this:

1. **Click your iPod in the Devices list.**

2. **Click the Video tab.**

3. **In the TV Shows section, select the Sync check box.**

4. **Choose an option from the drop-down menu.**

   - **All.** Choose this item to sync every TV show episode.

   - **1 Most Recent.** Choose this item to sync the most recent episode.

   - **3 Most Recent.** Choose this item to sync the three most recent episodes.

   - **5 Most Recent.** Choose this item to sync the five most recent episodes.

   - **10 Most Recent.** Choose this item to sync the ten most recent episodes.

   - **All unwatched.** Choose this item to sync all the episodes you haven't yet viewed.

   - **1 Most Recent Unwatched.** Choose this item to sync the most recent episode that you haven't yet viewed.

   - **3 Most Recent Unwatched.** Choose this item to sync the three most recent episodes that you haven't yet viewed.

- **5 Most Recent Unwatched.** Choose this item to sync the five most recent episodes that you haven't yet viewed.

- **10 Most Recent Unwatched.** Choose this item to sync the ten most recent episodes that you haven't yet viewed.

- **1 Least Recent Unwatched.** Choose this item to sync the oldest episode that you haven't yet viewed.

- **3 Least Recent Unwatched.** Choose this item to sync the three oldest episodes that you haven't yet viewed.

- **5 Least Recent Unwatched.** Choose this item to sync the five oldest episodes that you haven't yet viewed.

- **10 Least Recent Unwatched.** Choose this item to sync the ten oldest episodes that you haven't yet viewed.

Note A TV episode is unwatched if you haven't yet viewed it either in iTunes or your iPhone 3G. If you watch an episode on your iPhone 3G, the player sends this information to iTunes when you next sync.

5. **Select one of these options:**

   - **All TV shows.** Select this option to sync all your TV shows with your iPhone 3G.

   - **Selected.** Select this option to sync specific TV shows with your iPhone 3G. Choose either TV shows or Playlists in the menu, and then select the check boxes for the items you want to sync, as shown in figure 5.15.

6. **Click Apply.** iTunes syncs the iPhone 3G using your new TV show settings.

Genius To mark a TV episode as unwatched, in iTunes choose the TV Shows library, right-click (or Ctrl+Click on the Mac) the episode, and then choose Mark as New.

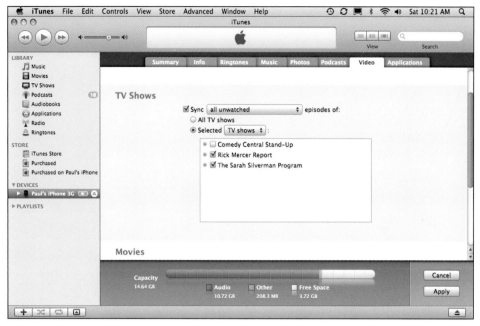

5.15 To sync specific TV shows, choose the Selected TV shows option, and then select the check boxes for each show you want synced.

Syncing computer photos to your iPhone 3G

No iPhone 3G's media collection is complete without a few choice photos to show off around the water cooler. One way to get those photos is to take them with your iPhone 3G's built-in digital camera. However, if you have some good pics on your computer, you can use iTunes to send those images to the iPhone 3G. Note that Apple supports a number of image file types — the usual TIFF and JPEG formats that you normally use for your photos as well as BMP, GIF, JPG2000 or JP2, PICT, PNG, PSD, and SGI.

If you use your computer to process lots of photos, and you want to take copies of some or all of those photos with you on your iPhone 3G, then follow these steps to get synced:

1. **Click your iPhone 3G in the Devices list.**

2. **Click the Photos tab.**

3. **Select the Sync photos from check box.**

4. **Choose an option from the drop-down menu:**

- **iPhoto (Mac only).** Choose this item to sync the photos, albums, and events you've set up in iPhoto.

- **Choose folder.** Choose this command to sync the images contained in a folder you specify.

- **Pictures (or My Pictures on Windows XP).** Choose this item to sync the images in your Pictures (or My Pictures) folder.

Note

If you have another photo-editing application installed on your computer, chances are it will also appear in the Sync photos from list.

5. **Select the photos you want to sync.** The controls you see depend on what you chose in Step 4:

- **If you chose either Pictures or the Choose folder.** In this case, select either the All photos option or the Selected folders option. If you select the latter, select the check box beside each subfolder you want to sync, as shown in figure 5.16.

- **If you chose iPhoto.** In this case, you get three further options: Select the All photos and albums option to sync your entire iPhoto library; select the *X* events option, where *X* is one of the following values that determines the number of iPhoto events that get synced: All, 1 most recent, 3 most recent, 5 most recent, 10 most recent, or 20 most recent; select the Selected albums option and then select the check box beside each album you want to sync.

6. **If you selected either the Selected folders option or the Selected albums option, use your mouse to click and drag the folders or albums to set the order you prefer.**

7. **Click Apply.** iTunes syncs the iPod using your new photo settings.

Note

iTunes doesn't sync exact copies of your photos to the iPhone 3G. Instead, it creates what Apple calls TV-quality versions of each image. These are copies of the images that have been reduced in size to match the iPhone 3G's screen size. This not only makes the sync go faster, but it also means the photos take up much less room on your iPhone 3G.

5.16 To sync photos from specific folders, choose the Selected folders option and then select the check boxes for each folder you want synced.

Syncing iPhone 3G photos to your computer

If you create a Safari bookmark on your iPhone 3G and then sync with your computer, that book-mark is transferred from the iPhone 3G to the default Web browser on your computer. That's a sweet deal that also applies to contacts and appointments, but unfortunately it doesn't apply to media files, which, with one exception, travel along a one-way street from your computer to your iPhone 3G.

Ah, but then there's that one exception, and it's a good one. If you take any photos using your iPhone 3G's built-in (and pretty good) camera, the sync process reverses itself and enables you to send some or all of those images to your computer. Sign us up!

Note

Actually, there's a second exception. If you use the iTunes application on your iPhone 3G to purchase or download music, those files are transferred to your computer dur-ing the next sync. iTunes creates a Store category called Purchased on *iPhone*, where *iPhone* is the name of your iPhone 3G. When the sync is complete, you can find your music there, as well as in the Music Library.

123

The iPhone 3G-to-computer sync process bypasses iTunes entirely. Instead, your computer deals directly with iPhone, and treats it just as though it was some garden-variety digital camera. How this works depends on whether your computer is a Mac or a Windows PC, so we'll use separate sets of steps.

To sync your iPhone 3G camera photos to your Mac, follow these steps:

1. **Connect your iPhone 3G to your Mac.** iPhoto opens, it adds your iPhone 3G to the Devices list, and it displays the photos from your iPhone 3G's Camera Roll album, as shown in figure 5.17.

Note   If you've imported some of your iPhone 3G photos in the past, you probably don't want to import them again. That's very sensible of you, and you can prevent that by hiding those photos. Select the Hide photos already imported check box.

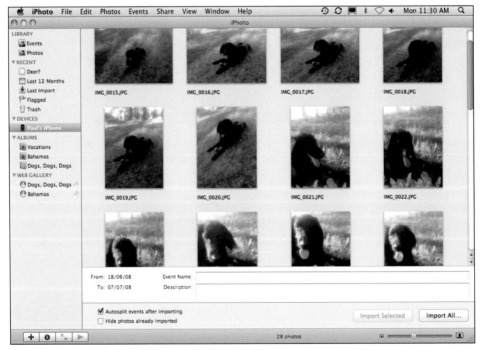

5.17 When you connect your iPhone 3G to your Mac, iPhoto shows up to handle the import of the photos.

2. **Use the Event Name text box to name the event that these photos represent.**

3. **Choose how you want to import the photos:**

   - If you want to import every photo, click Import All.

   - If you want to import only some of the photos, select the ones you want to import and then click Import Selected.

4. **If you want to leave the photos on your iPhone 3G, click Keep Originals.** Otherwise, click Delete Originals to clear the photos from your iPhone 3G.

Here's how things work if you're syncing with a Windows Vista PC:

1. **Connect your iPhone 3G to your Vista PC.** The AutoPlay dialog box appears, as shown in figure 5.18.

2. **Click Import pictures using Windows.** The rest of these steps assume you selected this option. However, if you have another photo management application installed, it should appear in the AutoPlay list, and you can click it to import the photos using that program.

3. **Type a tag for the photos.** A tag is a word or short phrase that identifies the photos.

5.18 When you connect your iPhone 3G to your Vista PC, the AutoPlay dialog box appears.

4. **Click Import.** Vista imports the photos and then opens Windows Photo Gallery to display them.

Finally, here's what happens if you're syncing with a Windows XP PC:

1. **Connect your iPhone 3G to your XP PC.** The Scanner and Camera Wizard opens, as shown in figure 5.19.

2. **Click Next.** The Choose Pictures to Copy dialog box appears.

3. **Select the check box associated with each photo you want to import and click Next.** The Picture Name and Destination dialog box appears.

4. **Type a name for the group of pictures, choose a location, and then click Next.** Windows XP imports the photos and asks you what you want to do next.

5. **Select Nothing, and click Next.**

6. **Click Finish.**

5.19 When you connect your iPhone 3G to your XP PC, the Scanner and Camera Wizard appears.

Preventing your iPhone 3G from sending photos to your computer

Each and every time you connect your iPhone 3G to your computer, you see iPhoto (on your Mac), the AutoPlay dialog box (in Windows Vista), or the Scanner and Camera Wizard (in Windows XP). This is certainly convenient if you actually want to send photos to your computer, but you might find that you do that only once in a blue moon. In that case, having to deal with iPhoto or a dialog box every time could cause even the most mild mannered among us to start pulling out our hair.

If you prefer to keep your hair, you can configure your computer to not pester you about getting photos from your iPhone 3G.

Note

Configuring your computer to not download photos from your iPhone 3G means that in the future you either need to reverse the setting to get photos or manually import your photos.

Here how you set this up on your Mac:

1. **Choose Finder ⇨ Applications to open the Applications folder.**

2. **Double-click Image Capture.** The Image Capture application opens.

3. **Choose Image Capture ⇨ Preferences.** The Image Capture Preferences window appears.

4. **Click the When a camera is connected, open menu, and then click No application, as shown in figure 5.20.**

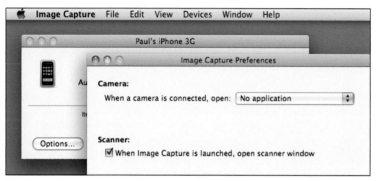

5.20 In the Image Capture Preferences window, choose No application to prevent iPhoto from starting when you connect your iPhone 3G.

5. **Choose Image Capture ⇨ Quit Image Capture.** Image Capture saves the new setting and then shuts down. The next time you connect your iPhone 3G, iPhoto ignores it.

Follow these steps to convince Windows Vista not to open the AutoPlay dialog box each time you connect your iPhone 3G:

1. **Choose Start ⇨ Default Programs to open the Default Programs window.**

2. **Click Change AutoPlay settings.** The AutoPlay dialog box appears.

3. **In the Devices section, open the Apple iPhone list and choose Take no action, as shown in figure 5.21.**

4. **Click Save.** Vista saves the new setting. The next time you connect your iPhone 3G, you won't be bothered by the AutoPlay dialog box.

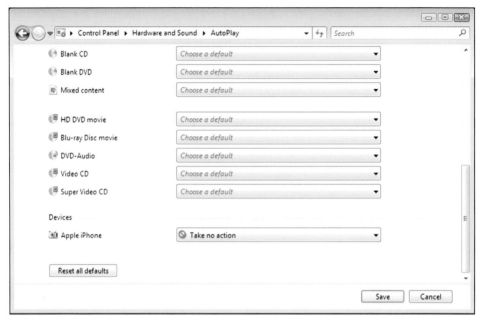

5.21 In the Apple iPhone list, choose Take no action to prevent the AutoPlay dialog box from appearing when you connect your iPhone 3G.

Follow these steps to configure Windows XP to bypass the Scanner and Camera Wizard when you connect your iPhone 3G:

1. **Connect your iPhone 3G to your Windows XP PC.**

2. **When the Scanner and Camera Wizard appears, click Cancel.**

3. **Choose Start ➪ My Computer to open the My Computer folder.**

4. **In the Scanners and Cameras section, right-click Apple iPhone and then click Properties.** The Apple iPhone 3G's Properties dialog box appears.

5. **Click the Events tab.**

6. **In the Select an event list, choose Camera connected.**

7. **Select the Take no action option, as shown in figure 5.22.**

5.22 In the Apple iPhone Properties dialog box, choose Take no action to prevent the Scanner and Camera Wizard from appearing when you connect your iPhone 3G.

8. **Click OK.** XP saves the new setting. The next time you connect your iPhone 3G, the Scanner and Camera Wizard remains offstage.

Syncing media with two or more computers

It's a major drag, but you can't sync the same type of content to your iPhone 3G from more than one computer. For example, suppose you're syncing photos from your desktop computer. If you then connect your iPhone 3G to another computer (your notebook, for example), crank up iTunes, and then select the Sync photos from check box, iTunes coughs

5.23 Syncing the same type of content from two different computers is a no-no in the iTunes world.

up the dialog box in figure 5.23. As you can see, iTunes is telling you that if you go ahead with the photo sync on this computer, it will blow away all your existing iPhone 3G photos and albums!

So there's no chance of syncing the same iPhone 3G with two different computers, right? Not so fast, my friend! Let's try another thought experiment. Suppose you're syncing your iPhone 3G with your desktop computer, but you're not syncing Movies. Once again, you connect your iPhone 3G to your notebook computer (or whatever), crank up iTunes, and then select the Sync movies check box. Hey, no ominous warning dialog box! What gives?

The deal here is that if iTunes sees that you don't have any examples of a particular type of content (such as movies) on your iPhone 3G, it lets you sync that type of content, no questions asked.

In other words, you *can* sync your iPhone 3G with multiple computers, although in a roundabout kind of way. The secret is to have no overlapping content types on the various computers you use for the syncing. For example, let's say you have a home desktop computer, a notebook computer, and a work desktop computer. Here's a sample scenario for syncing your iPhone 3G with all three machines:

- **Home desktop (music and video only).** Select the Sync music check box in the Music tab, and select all the Sync check boxes in the Video tab. Deselect the Sync check boxes on the Photos and Podcasts tabs.

- **Notebook (photos only).** Select the Sync photos from check box on the Photos tab. Deselect all the Sync check boxes in the Music, Podcasts, and Video tabs.

- **Work desktop (podcasts only).** Select the Sync box in the Podcasts tab. Deselect the Sync check boxes in the Music, Photos, and Video tabs.

129

How Can I Get More Out of My iPhone 3G's Media Features?

Your iPhone 3G is a perk-filled device, to be sure, and one of the best of those perks is that the iPhone 3G moonlights as a digital media player. You've got music, movies, TV shows, podcasts, audiobooks, photos, and even YouTube videos right there in the palm of your hand. So when you're tired of calling, researching, e-mailing, scheduling, and other serious iPhone 3G pursuits, you can kick back with a tune or a show to relax. However, your iPhone 3G is capable of more than just playing and viewing media. It's actually loaded with cool features that enable you to manipulate media files, and use those files to enhance other parts of your digital life. This chapter is your guide to these features.

Creating a Custom Menu Bar for the iPhone 3G's iPod

Your iPhone 3G is a living, breathing iPod, thanks to its built-in iPod application, which you can fire up any time you want by tapping the iPod icon in the Home screen's menu bar. Of course, your iPhone 3G doesn't come with the famous click-wheel found on physical iPods, so you need some other way of getting around. The iPhone 3G's solution is to preset a series of *browse buttons*, each one of which represents a collection of media files organized in some way. For example, tapping the Songs browse button displays a list of all the songs on your iPhone 3G.

You see four browse buttons in the default menu bar — Playlists, Artists, Songs, and Videos — and a fifth button called More that displays a list of six more browse buttons. Here's a summary of all ten buttons:

- **Playlists.** This browse button displays your playlists, which are collections of songs that you (or iTunes) have gathered together to play as a group. Your iPhone 3G also comes with a special playlist called On-The-Go that we talk about a bit later. Tap the playlist to see the songs it holds.

- **Artists.** This browse button displays an alphabetical list of the artists who perform all the songs in your iPhone 3G's music collection. Tap the artist to see an album or songs.

- **Songs.** This browse button offers an alphabetical list of every song on your iPhone 3G. Tap a song to crank it up.

- **Videos.** This browse button displays a list of the videos you have imported to your iPhone 3G. They're organized by category: Movies, Movie Rentals, TV Shows, Music Videos, and Video Podcasts. Tap a video to start playing it.

- **Albums.** This browse button gives you an alphabetical list of your iPhone 3G's album collection, and with each album you see the title, artist, and album art. Tap the album to see the songs it holds.

- **Audiobooks.** This browse button sends in a list of the audiobooks that you have imported to your iPhone 3G. Tap a book to launch the story.

- **Compilations.** This browse button offers a list of your iPhone 3G albums that contain multiple artists, such as soundtracks and collections of Christmas music. Tap an album to see its tracks.

- **Composers.** This browse button displays an alphabetical list of the composers who wrote all the songs in your iPhone 3G's music library. Tap the composer to see his or her albums or songs.

Genius   Sometimes iTunes goes a bit haywire and classifies a single-artist album as a compilation. If you see such an album in your 3G's Compilations list, open iTunes on your computer, select all the tracks on that album, choose File ⇨ Get Info, choose No in the Compilation list, and then click OK. The next time you sync your iPhone 3G, that album won't appear in Compilations.

- **Genres.** This browse button gives you a list of the genres represented by your iPhone 3G's music (Pop, Rock, Classical, Jazz, and so on). Tap the genre to see its composers.

- **Podcasts.** This browse button offers a list of all the podcasts that you have stored on your iPhone 3G, organized alphabetically by publisher. A blue dot next to a publisher helpfully tells you that there are episodes of that podcast you haven't listened to. Tap the publisher to see the individual podcast episodes, sorted by date. Those episodes you haven't gotten around to yet have a blue dot to their left.

With ten browse buttons available but only four menu bar slots, chances are there are one or more buttons that you use frequently but are exiled to the More list, thus requiring an extra tap to access. Extra taps are uncool! So if there's a browse button on the More list that you use all the time, you can move it to the menu bar for easier access. Here's how:

1. **On the Home screen, tap iPod to open the iPod application.**

2. **Tap More in the menu bar.**

3. **Tap Edit.** Your iPhone 3G displays the Configure screen, which shows all ten browse buttons, as shown in figure 6.1.

4. **Drag a browse button that you want to add to the menu bar and drop it on whatever existing menu bar browse button you want it to replace.** For example, if you want to replace the Videos browse button with Podcasts, drag the Podcasts button and drop it on Videos. Your iPhone 3G replaces the old browse button with the new one.

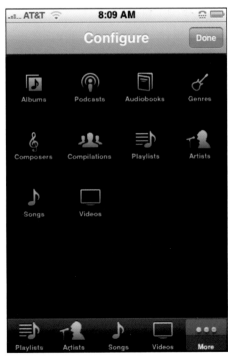

6.1 Use the Configure screen to create a custom iPod menu bar.

5. **Repeat step 4 to add any of your other preferred browse buttons to the menu bar.**

6. **Tap Done to save the new menu bar configuration.**

Getting More Out of Your iPhone 3G's Audio Features

The iPod application on your iPhone 3G is built with audio in mind, and it lets you crank music, music videos, audiobooks, and podcasts. If you have a fast Wi-Fi connection going (a 3G cellular connection will do in a pinch), you can even use your iPhone 3G to purchase music directly from the iTunes Store (tap the iTunes icon in the Home screen). Playing the track you want is a snap on your iPhone 3G: Tap iPod, tap a browse button, locate the track, and then tap it. However, your iPhone 3G is more than a simple tap-and-play device, and the following sections show you how to take advantage of some of the phone's more useful audio features.

Genius If you use your iPhone 3G's iPod feature all the time, go ahead and customize the Home button to launch iPod. Tap Settings, tap General, tap Home Button to open the Home Button screen, and then tap iPod. Now you can switch to the iPod super-fast by double-clicking the Home button from any screen.

Using audio accessories with your iPhone 3G

As soon as the original iPhone was announced, a rather large cottage industry of iPhone accessories formed, seemingly overnight. Suddenly the world was awash in headsets (wired and Bluetooth), external speakers, FM transmitters, and all manner of cases, car kits, cables, and cradles. There are places that sell iPhone 3G accessories scattered all over the Web, but the following sites are faves with us:

- **Apple.** http://store.apple.com/us/browse/home/shop_iphone
- **Belkin.** www.belkin.com/ipod/iphone/
- **Griffin.** www.griffintechnology.com/devices/iphone/
- **NewEgg.** www.newegg.com/
- **EverythingiCafe.** http://store.everythingicafe.com/

Here are a few notes to remember when shopping for and using audio-related accessories for your iPhone 3G:

- **Look for the logo.** Your iPhone 3G is an iPod in fancy phone clothes. It's a completely different device that doesn't fit or work with many iPod accessories. To be sure what you're buying is iPhone-friendly, look for the "Works with iPhone" logo.

- **Headsets, headphones, and earpieces.** For our money (literally), one of the best and most welcome design improvements in the iPhone 3G involves the headset jack, which is no longer annoyingly recessed into the case (as it was on the original iPhone). This means that just about any headset that uses a garden-variety stereo mini-plug will fit your iPhone 3G without a hitch, and without requiring the purchase of an adapter. Yes!

- **External speakers.** There are legions of external speakers made for the iPod where you simply dock the iPod in the device and wail away. Unfortunately, the dimensions of the iPhone 3G's bottom panel are different than any of the iPod models, so you won't be able to just plug-and-play your iPhone 3G. Instead, you need an adapter — such as Apple's Universal Dock Adapter — to ensure a proper fit.

- **FM transmitters.** These are must-have accessories for car trips because they send the iPhone 3G's output to an FM station, which you then play through your car stereo. The FM transmitters that work with the iPod don't generally work with iPhone 3Gs, so look for one that's designed for the iPhone 3G.

- **Electronic interference.** Because your iPhone 3G is, after all, a phone, it generates a nice little field of electronic interference, which is why you need to switch it to Airplane mode when you're flying (see Chapter 2). That same interference can also wreak havoc on nearby external speakers and FM transmitters, so if you hear static when playing audio, switch to Airplane mode to get rid of it.

Creating a favorite tunes playlist for your iPhone 3G

Your iTunes library includes a Rating field that enables you to supply a rating for your tracks: one star for songs you don't like so much, up to five stars for your favorite tunes. You click the song you want to rate, and then click a dot in the Rating column (click the first dot for a one-star rating, the second dot for a two-star rating, and so on). Rating songs is useful because it enables you to organize your music. For example, the Playlists section includes a My Top Rated playlist that includes all your four- and five-star-rated tunes, ordered by the Rating value.

Rating tracks comes in particularly handy when deciding which music to use to populate your iPhone 3G. If you have tens of gigabytes of tunes, only some of them will fit on your iPhone 3G. How do you choose? In Chapter 5, we show you that you can organize your music into playlists, and then sync the playlists you want to hear on your iPhone 3G. Another possibility would be to rate your songs, and then just sync the My Top Rated playlist to your iPhone 3G.

The problem with the My Top Rated playlist is that it includes only your four- and five-star-rated tunes. You can fit thousands of tracks on your iPhone 3G, but it's unlikely that you've got thousands of songs rated at four stars or better. To fill out your playlist, you should also include songs rated at three stars, a rating that should include lots of good, solid tunes.

To set this up, you have two choices:

- **Modify the My Top Rated playlist.** Right-click (or Ctrl+Click on a Mac) the My Top Rated playlist, and then click Edit Playlist. In the Smart Playlist dialog box, click the second star, and then click OK.

- **Create a new playlist.** This is the way to go if you want to leave My Top Rated as your best music. Choose File ⇨ New Smart Playlist to open the Smart Playlist dialog box. Choose Rating in the Field list, Is Greater Than in the Operator list, and then click the second star. Figure 6.2 shows the configured dialog box. Click OK, type a title for the playlist (such as Favorite Tunes), and then press Return (or Enter).

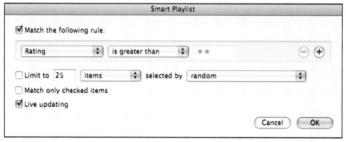

6.2 Use the Smart Playlist dialog box to create a playlist that contains your tracks rated at three stars or more.

The next time you sync your iPhone 3G, be sure to include either the My Top Rated playlist or the Smart Playlist you created.

Rating a song on your iPhone 3G

If you use song ratings to organize your tunes, you might have come across some situations where you'd like to rate a song that's playing on your iPhone 3G:

- You used your iPhone 3G to download some music from the iTunes Store, and you want to rate that music.

- You're listening to a song on your iPhone 3G and decide that you've given a rating that's either too high or too low and you want to change it.

In the first case, you could sync the music to your computer and rate it there; in the second case, you could modify the rating on your computer and then sync with your iPhone 3G. However, these solutions are lame because you have to wait until you connect your iPhone 3G to your computer. If you're out and about, you want to rate the song *now*, while it's fresh in your mind.

Yes, you can do that with your iPhone 3G:

1. **Locate the song you want to rate and tap it to start the playback.** Your iPhone 3G displays the album art and the name of the artist, song, and album at the top of the screen. To the right of these names, iPhone 3G displays the Details icon.

2. **Tap the Details icon in the upper-right corner of the screen (just below the battery status).** Your iPhone 3G "turns" the album art and displays a list of the songs on the album. Above that list are the five rating dots.

3. **Tap the dot that corresponds to the rating you want to give the song.** For example, to give the song a four-star rating, tap the fourth dot from the left, as shown in figure 6.3.

4. **Tap the album art icon in the upper-right corner.** Your iPhone 3G saves the rating and returns you to the album art view.

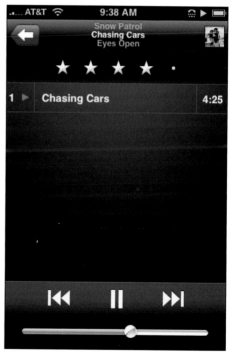

6.3 Tap the dot that corresponds to the rating you want to give the currently playing track.

The next time you sync your iPhone 3G with your computer, iTunes notes your new ratings and applies them to the same tracks in the iTunes library.

Browsing iPhone 3G music with Cover Flow

Here in the first decade of the twenty-first century, physical CD collections are suffering the same fate that vinyl LP collections went through in the 1980s: they're disappearing. We're not crying in our beer over this trend because a large anthology of CDs is an eyesore. However, there's one thing we do miss: flipping through CD covers looking for something that catches our eye. You may not be able to do this anymore, but your iPhone 3G gives you the next best thing: Cover Flow. This feature displays all your album art as a kind of flip book. Only in this case we should call it a "flick" book because you can use your finger to flick back and forth through the album art, just like flipping through a CD collection.

Switching to Cover Flow mode takes a mere two steps:

1. **In the Home screen, tap the iPod icon to open the iPod application if it's not already open.**

2. **Rotate your iPhone 3G into the landscape position.** Your iPhone 3G switches to the Cover Flow view, and displays either the first album in your collection, or the album associated with whatever's currently playing (or most recently played), as shown in figure 6.4.

6.4 Open iPod and rotate your iPhone 3G into landscape mode to see the Cover Flow view.

You can flick to the right and left to navigate the albums. To see the tracks on an album, tap the album art (or the "i" icon in the bottom-right corner). To return to the regular iPod view, rotate your iPhone 3G into the upright position.

Answering an incoming call while listening to music on the headset

If you're listening to music on your iPhone 3G and a call comes in, you obviously don't want the caller to be subjected to Death Cab for Cutie at top volume. Fortunately, your iPhone 3G smartphone is smart enough to know this, and it automatically pauses the music. If you have your iPhone 3G headset on when the call arrives, use the following techniques to deal with it:

- **Answer the call.** Press and release the headset's mic button (it's the plastic button on one of the headset's cords).

- **Decline the call (send it directly to voicemail).** Press and hold the mic button for about two seconds, and then release. If you hear a couple of beeps, you successfully declined the call.

- **End the call.** Press and release the mic button.

Creating a custom ringtone for your iPhone 3G

Your iPhone 3G comes stocked with 25 predefined ringtones. Although some of them are amazingly annoying, you ought to be able to find one you can live with. If you can't, or if you crave something unique, you can create a custom ringtone and use that.

The easiest way to cobble together a custom ringtone is to convert a song you purchase through iTunes:

1. **In iTunes, select the track you want to use.** If you're not sure which tracks in your library are ringtone-friendly, look for the bell icon in the Ringtone column (which itself shows a bell icon in the column header).

Note

No Ringtone column in sight? No problem. Right-click (or Ctrl+Click on your Mac) any column header, and then click Ringtone.

2. **Choose Store ⇨ Create Ringtone.** The Ringtone Editor appears. If you run the Create Ringtone command and an error message appears telling you that iTunes can't connect to the iTunes Store, it means your country's version of the iTunes Store can't handle ringtone purchases. That's a drag, but we show you a couple of ways to work around it in the next set of steps.

3. **Click and drag the highlighted area to specify which part of the song you want to use for your ringtone.** The maximum size of the ringtone snippet is 30 seconds.

4. **In the Looping pop-up menu, choose the interval you want between rings.** Click Preview any time you want to hear your ringtone in action.

5. **Click Buy.** iTunes rings up the purchase and adds the new ringtone to its Ringtones category.

The Create Ringtone feature is easy, for sure, but it only works on certain songs, and many international versions of the iTunes Store don't offer support for purchasing ringtones. Fortunately, you can get around both limitations using GarageBand, Apple's application for making homebrew music.

First, here are the steps to follow to create a ringtone out of any song in your iTunes library:

1. **Click the GarageBand icon in the Dock, click Create New Music Project, and then click Create.** GarageBand starts a new project for you.

2. **Choose Track ⇨ Delete Track to get rid of the default piano track.**

3. **Switch to iTunes, click and drag the song you want to use for your ringtone, and drop it inside GarageBand.** The program creates a new track for the song.

4. **Click the Cycle Region button.** GarageBand adds the Cycle Region tool above the track, as shown in figure 6.5. The Cycle Region defines the portion of the song that you'll use for the ringtone.

5. **Click and drag the Cycle Region to the approximate area of the song you want to use for the ringtone.**

6. **Click and drag the left edge of the Cycle Region to define the starting point of the ringtone.**

7. **Click and drag the right edge of the Cycle Region to define the starting point of the ringtone.**

Note

The maximum length for a GarageBand ringtone is 40 seconds. To see how long the Cycle Region is, choose Control ⇨ Show Time in LCD.

8. **Choose File ⇨ Save As, type a name for the ringtone, and then click Save.**

Cycle Region button   Cycle Region tool

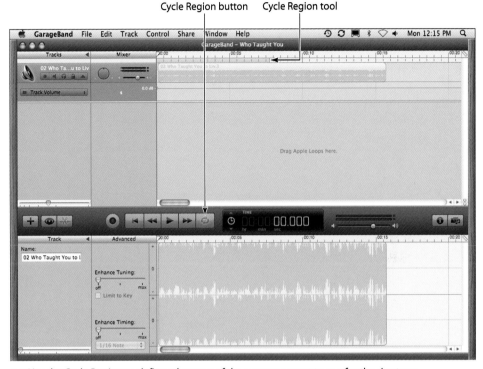

6.5 Use the Cycle Region to define what part of the song you want to use for the ringtone.

9. **Choose Share ▷ Send Ringtone to iTunes.** GarageBand converts the track to a ringtone, and then adds it to the Ringtones category in iTunes.

Genius There's no reason you have to use commercial music for your ringtone. GarageBand makes it easy to create your own music from scratch. For example, choose File ▷ New, click Magic GarageBand, and then click a music genre. GarageBand creates a whole song for you, and you can even add your own instruments! (Click Audition, and then click Create Project when you're done.)

The next time you sync your iPhone 3G, click the Ringtones tab, select the Sync ringtones check box, choose your custom ringtone, and then click Apply. To apply the custom ringtone on your iPhone 3G, tap Settings in the Home screen, tap Sounds, and tap Ringtone. Your ringtone should appear in the Custom section of the Ringtone screen. Tap it to use the snippet as your ringtone.

Creating a playlist on your iPhone 3G

The playlists on your iPhone 3G are those you've synced via iTunes, and those playlists are either generated automatically by iTunes or they're ones you've cobbled together yourself. However, when you're out in the world and listening to music, you might come up with an idea for a different collection of songs. It might be female singers, boy bands, or songs with animals in the title.

Whatever your inspiration, don't do it the hard way by picking out and listening to each song one at a time. Instead, you can use your iPhone 3G to create a playlist on the fly or, really, on the go, because your phone has a special playlist called On-The-Go that you create yourself, right on your iPhone 3G.

To create an On-The-Go playlist, follow these steps:

1. **Open the iPod application.**

2. **Tap the Playlists icon.** This displays your playlists.

3. **Tap the On-The-Go playlist.** A list of all of your songs appears. You can also click one of the browse buttons to help find your music.

4. **Scroll through the list and tap the blue + key next to each song you want to add to your list.** Your iPhone 3G turns a song gray when you add it, as shown in figure 6.6.

5. **Tap Done.** Your iPhone 3G displays the On-The-Go playlist.

6. **To add more tracks, tap the + button in the upper-left corner, select another playlist, and then repeat Step 4 for each song you want to add.**

7. **In the On-The-Go playlist screen, tap Done.** Your playlist is ready to play.

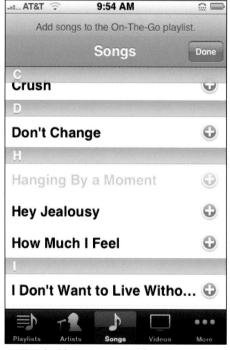

6.6 Tap the blue + icons to add songs to your On-The-Go playlist.

Your On-The-Go playlist isn't set in stone by any means. You can get rid of songs, change the song order, and add more songs. Follow these steps:

1. **In the iPod application, tap the Playlists icon to see your playlists.**

2. **Tap On-The-Go.** Your iPhone 3G sends in your On-The-Go playlist.

3. **Tap Edit.** This changes the list to the editable version, as shown in figure 6.7.

4. **To remove a song, tap the red Delete icon to the left of the song, and then tap the Delete button that appears.** If you change your mind, tap the red Delete icon on the left to cancel the deletion.

5. **To move a song within the playlist, slide the song's drag icon (it's on the right) up or down to the position you prefer.**

6. **To add more tracks, tap the + button in the upper-left corner, select another playlist, and then tap the blue + key next to each song you want to add.**

7. **When you are done editing, tap Done.** This sets the playlist.

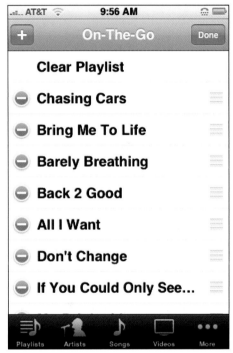

6.7 The On-The-Go playlist in Edit mode.

Note

If your On-The-Go playlist is a bit of a mess, or if your mood suddenly changes, don't delete all the tracks one-by-one. Instead, open the On-The-Go playlist, tap Edit, and then tap Clear Playlist. When your iPhone 3G asks you to confirm, tap Clear Playlist.

Customizing your iPhone 3G's audio settings

Audiophiles in the crowd don't get much to fiddle with in the iPhone 3G, but there are a few audio settings to play with. Here's how to get at them:

1. **Press the Home button to get to the Home screen.**

143

2. **Tap the Settings icon.** This opens the Settings screen.

3. **Tap the iPod icon.** Your iPhone 3G displays the iPod settings screen, as shown in figure 6.8.

You get four settings to try out:

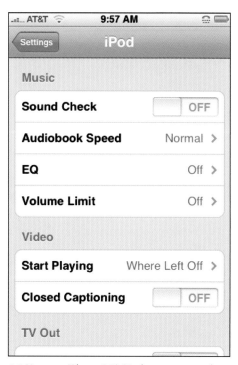

- **Sound Check.** Every track is recorded at different audio levels, so invariably you get some tracks that are louder than others. With the Sound Check feature, you can set your iPhone 3G to play all of your songs at the same level. This feature only affects the baseline level of the music and doesn't change any of the other levels, so you still get the highs and lows. If you use it, you don't need to worry about having to quickly turn down the volume when a really loud song comes on. To turn on Sound Check, in the iPod settings page, tap the Sound Check switch to the On position.

6.8 Use your iPhone 3G's iPod screen to muck around with the audio settings.

- **Audiobook Speed.** You use this setting to change the speed at which your iPhone 3G plays audiobooks. If the narrator is reading too fast you can slow him down or the other way around. The feature does a surprisingly good job of keeping the speed change from raising or lowering the voice too much, although there is a noticeable difference. In the iPod settings page, tap Audiobook Speed, and then tap the speed you want: Slower, Normal (the default), or Faster.

- **EQ.** This setting controls your iPhone 3G's built-in equalizer, which is actually a long list of preset frequency levels that affect the audio output. Each preset is designed for a specific type of audio: vocals, talk radio, classical music, rock, hip-hop, and lots more. To set the equalizer, tap EQ and then tap the preset you want to use (or tap None to turn off the equalizer).

- **Volume Limit.** You use this setting to prevent the iPhone 3G's volume from being turned up too high and damaging your (or someone else's) hearing. You know, of course, that pumping up the volume while you've got your earbuds on is an audio no-no, right? We thought so. However, we also know that when a great tune comes on, it's often a little too tempting to go for 11 on the volume scale. If you can't resist the temptation, use Volume Limit to limit the damage. Tap Volume Limit and then drag the Volume slider to maximum allowed volume.

Genius

If you're setting up an iPhone for a younger person, you should set the Volume Limit. However, what prevents the young whippersnapper from setting a higher limit? You can. In the Volume Limit screen, tap Lock Limit Volume. In the Set Code screen, tap out a four-digit code, and then tap the code again to confirm. This disables the Volume slider in the Volume Limit screen.

Getting More Out of Your iPhone 3G's Video Features

The iPhone 3G is a visual medium. It uses a touchscreen after all. The large screen makes it perfect as a portable media device. Its iPod application organizes and plays back videos that you import from your computer. Your commute (okay, your *nondriving* commute) doesn't have to be boring any more. Just put in your headphones and start an episode of your favorite show.

Playing videos, movies, and TV shows

In an ideal world, you'd watch all of your videos on a comfy couch in front of a flat-screen TV with a rockin' surround-sound system. Unfortunately, all that equipment isn't exactly portable. However, your iPhone 3G is *very* portable and what's more, you probably carry it with you just about everywhere. Throw that nice large screen into the mix, and you've got yourself a great portable video player.

To watch a video on your iPhone 3G, follow these steps:

1. **Tap the iPod icon on the Home screen.** This opens the iPod application.

2. **Tap the Videos browse button.** Your iPhone 3G displays a list of your videos, organized by type: Rented Movies, Movies, TV Shows, and Music Videos.

3. **Tap the video you want to watch.** The video opens on the screen and starts playing.

4. **Turn the screen to the landscape position to watch the video.**

When you first see an iPhone 3G video, you might think you have no way to control the playback because there are no controls in sight. Fortunately for you, Apple realized that watching a movie with a bunch of buttons pasted on the screen wouldn't exactly enhance the movie-watching experience. We agree. The buttons are actually hidden, but you can force them out of hiding by tapping the screen. Figure 6.9 shows what you see.

6.9 Tap the video to reveal the playback controls.

Here's what you see:

- **Progress Bar.** This bar shows you where you are in the video playback. The white ball shows you the current position, and you can drag the ball left (to rewind) or right (to fast forward). To the right is the time remaining in the video and on the left is the time elapsed.

- **Fill/Fit the Screen.** This button in the upper-right corner toggles the video between filling the entire screen (which may crop the outside edges of the video) and fitting the video to the screen width (which gives you letterboxed video with black bars above and below the video).

Note
You can also switch between filling the screen and fitting the screen by double-tapping the screen.

- **Previous.** Tap this button (the left-pointing arrows) to return to the beginning of the video or, if you're already at the beginning, to jump to the previous chapter (if the video has multiple chapters, as do most movies). Tap and hold this button to rewind the video.

- **Next.** Tap this button (the right-pointing arrows) to jump to the next chapter (if the video has multiple chapters). Tap and hold this button to fast-forward the video.

- **Pause/Play.** Tap this button to pause the playback, and then tap it again to resume.

- **Chapter Guide.** Tap this button to see a list of chapters in the video. (You don't see this button if the video doesn't have multiple chapters).

Note

You can use your iPhone 3G headset to control video playback. Click the mic button once to play or pause. Click it twice to skip to the next chapter.

- **Volume Bar.** This bar controls the video volume level. Drag the white ball to change the level. You can also use volume controls on the side of the phone.

- **Done.** Tap this button to stop the video and return to the list of videos on your iPhone 3G. You can also press the Home button to stop the video and wind up on the Home screen.

Note

If you stop the video before the end, the next time you tap the video, it resumes the playback from the spot where you stopped it earlier. Nice!

Playing just the audio portion of a music video

When you play a music video, you get a two-for-one media deal: great music and (hopefully) a creative video. That's nice, but the only problem is you can't separate the two. For example, sometimes it might be nice to listen to just the audio portion of the music video. Why? Because you can't do anything else on your iPhone 3G while a video is playing. If you press the Home button, for example, the video stops and the Home screen appears. That certainly makes sense, so it would be nice to be able to play just the audio portion, because your iPhone 3G *does* let you perform some other tasks while playing audio.

Unfortunately, your iPhone 3G doesn't give you any direct way to do this. You might think your only hope is to rip or purchase the song separately, but we've figured out a workaround. The secret is that if you add a music video to a regular music playlist, iPhone 3G treats the music video like a regular song. When you then play it on your iPhone 3G using that playlist, you hear just the audio portion (and see just the first frame of the video as the album art).

To add a music video to a playlist in iTunes, follow these steps:

1. **Open iTunes on your computer.**
2. **It's best to use a custom playlist for this, so create your own playlist if you haven't done so already.**
3. **In Playlists, click Music Videos.** A list of all your music videos appears.
4. **Right-click (or Ctrl+Click on a Mac) the video you want to work with, click Add to Playlist, and then click a playlist.** iTunes adds the music video to the playlist.

5. Repeat Step 4 for any other music videos you want to just listen to.

Sync your iPhone 3G to download the updated playlist. Then tap iPod, Playlists, the playlist you used, and then the music video. Your iPhone 3G plays the audio portion, and displays the first frame of the video, as shown in figure 6.10. You're now free to move about the iPhone 3G cabin while listening to the tune.

Playing iPhone 3G videos on your TV

You can carry a bunch of videos with you on your iPhone 3G so why shouldn't you be able to play them on a TV if you want? Well, you can. You have to buy another cord, but that's the only investment you have to make to watch iPhone 3G videos right on your TV.

To hook your iPhone 3G up to your TV, you have three choices:

6.10 When you play a music video via a regular playlist, your iPhone 3G treats it like a regular audio track.

- **Apple iPod AV cable.** This $19 cable has a stereo mini-plug on one end that connects to the iPhone 3G's headset jack, and it has RCA connectors on the other end that connect to the AV inputs of your TV.

- **Apple Component AV cable.** This $49 cable has a dock connector on one end that plugs into the iPhone 3G's dock connector, and it has component connectors on the other end that connect to the component inputs of your TV.

- **Apple Composite AV cable.** This $49 cable has a dock connector on one end that plugs into the iPhone 3G's dock connector, and it has composite connectors on the other end that connect to the composite inputs of your TV.

The cable you choose depends on the type of TV you have. Older sets have AV inputs or possible composite inputs, while most newer flat-screen TVs have component inputs.

After setting up your cables, set your TV to the input and play your videos as you normally would.

Note

Your iPhone 3G offers a couple of settings that affect the TV output. See the next section to learn more about them.

Customizing your iPhone 3G's Video settings

Your iPhone 3G offers a few video-related settings that you can try on for size. Follow these steps to get at them:

1. **Press the Home button to get to the Home screen.**

2. **Tap the Settings icon.** This opens the Settings screen.

3. **Tap the iPod icon.** Your iPhone 3G displays the iPod settings screen.

You get four settings to meddle with:

● **Start Playing.** This setting controls what your iPhone 3G does when you stop and restart a video. You have two choices: Where Left Off (the default), which picks up the video from the same point where you stopped it; and From Beginning, which always restarts the video from scratch. Tap Start Playing, and then tap the setting you prefer.

● **Closed Captioning.** This setting toggles support for closed captioning on and off, when it's available. To turn on this feature, tap the Closed Captioning switch to the On position.

● **Widescreen.** This setting toggles support for widescreen TV output. If you have a widescreen TV and you want to play iPhone 3G videos on the set, tap the Widescreen switch to the On position.

● **TV Signal.** This setting specifies the TV output signal. If you're going to play videos on a TV, tap TV Signal and then tap either NTSC or PAL.

Getting More Out of Your iPhone 3G's Photos

Your iPhone 3G comes with a built-in digital camera so you can take pictures while you're running around town. The camera on the iPhone 3G is pretty basic: It's a 2-megapixel camera with no scene modes, no zoom, and no video. It produces a decent image if you have plenty of available light and the subject isn't moving, but otherwise the photos it produces are on the mediocre side.

While the camera itself may not win any awards, what you can do with photos on your iPhone 3G is pretty cool. You can e-mail photos to friends, take a photo and assign it to a contact, or make a slide show with music in the background. Its large screen makes it the perfect portable photo album. No more whipping out wallet shots of your kids. Now you can show people your iPhone 3G photo album!

To get to your photos, tap the Photos icon in your iPhone 3G's Home screen to display the Photo Albums screen, which lists your photo albums. Tap an album to see its pictures, and then tap the picture you want to check out.

Scrolling, rotating, zooming, and panning photos

You can do quite a lot with your photos once they are in your iPhone 3G, and it isn't your normal photo browsing experience. You aren't just a passive viewer because you can actually take some control over what you see and how the pictures are presented.

You can use the following techniques to navigate and manipulate your photos:

- **Scroll.** You move forward or backward through your photos by flicking. Flick from the right to left to view the next photo; flick left to right to view the previous shot. Alternatively, tap the screen, and then tap the Previous and Next buttons to navigate your photos.

- **Rotate.** When a landscape shot shows up on your iPhone 3G it gets letterboxed at the top (that is, you see black space above and below the image). To get a better view, rotate the screen into the landscape position and the photo rotates right along with it, filling the entire screen. When you come upon a photo with a portrait orientation, rotate the iPhone 3G back to the upright position for best viewing.

- **Zoom.** Zooming magnifies the shot that's on the screen. There are two methods to do this.

  - **Double-tap the area of the photo that you want to zoom in on.** The iPhone 3G doubles the size of the portion you tapped. Double-tap again to return the photo to its original size.

  - **Spread and pinch.** To zoom in, spread two fingers apart over the area you want magnified. To zoom back out, pinch two fingers together.

- **Pan.** After you zoom in on the photo, you may find that the iPhone 3G didn't zoom in exactly where you want or you may just want to see another part of the photo. Drag your finger across the screen to move the photo along with your finger, an action known as *panning*.

Note You can scroll to another photo if you're zoomed in, but it takes a lot more work to get there because the iPhone 3G thinks you're trying to pan. For faster scrolling, return the photo to its normal size, and then scroll.

Adding an existing photo to a contact

You can assign a photo from one of your albums to any of your contacts. This is one of our favorite iPhone 3G features because it means that when the person calls you, his or her smiling mug appears on your screen. Now *that's* caller ID! There are two ways to assign a photo to a contact: You can assign the photo straight from a photo album, or you can go through the Contacts list.

First, here's how you assign a photo from a photo album:

1. **Tap Photos in the Home screen.**

2. **Tap the photo album that has the image you want to use.**

3. **Tap the photo you want to use.** Your iPhone 3G opens the photo.

4. **Tap the image to reveal the controls.**

5. **Tap the Action button.** The Action button is the button on the left side of the menu bar. (If you don't see the menu bar, tap the screen.) iPhone 3G displays a list of actions you can perform.

6. **Tap Assign To Contact.** A list of all of your contacts appears.

7. **Tap the contact you want to associate with the photo.** The Move and Scale screen appears.

8. **Drag the image so that it's positioned on the screen the way you want.**

9. **Pinch or spread your fingers over the image to set the zoom level you want.**

10. **Tap Set Photo.** iPhone 3G assigns the photo to the contact and returns you to your photo album.

To assign a photo using the Contacts list, follow these steps:

1. **On the Home screen, tap the Contacts icon to open the Contacts list.**

2. **Tap the contact that you want to add a photo to.** Your iPhone 3G displays the contact's Info screen.

3. **Tap Edit to put the contact into Edit mode.**

4. **Tap Add Photo.** iPhone 3G displays a list of photo options.

5. **Tap Choose Existing Photo.** Your iPhone 3G displays the Photo Albums screen.

6. **Tap the album that contains the photo you want to use.**

7. **Tap the photo you want.** The Move and Scale screen appears.

8. **Drag the image so that it's positioned on the screen the way you want.**

9. **Pinch or spread your fingers over the image to set the zoom level you want.**

10. **Tap Set Photo.** iPhone 3G assigns the photo to the contact and returns you to the Info screen.

11. **Tap Done.** Your iPhone 3G exits Edit mode.

Taking a contact's photo with the iPhone 3G camera

If you don't have a picture of a contact handy, that's not a problem because you can take advantage of your iPhone 3G's camera to snap his or her image the next time you get together. You can do this either using the Camera application or via the Contacts list.

To assign a photo from the Camera application, follow these steps:

1. **In the Home screen, tap the Camera icon on the Home screen to enter the Camera application.** A shutter appears on the screen.

2. **Frame the person on your screen and say "Okay, say iPhone 3Geeeeee."**

3. **Tap the Camera key at the bottom of the screen to snap the picture.**

4. **Tap the Camera Roll icon in the bottom-left corner.** This opens the Camera Roll screen.

5. **Tap the photo you just took.** Your iPhone 3G opens the photo and reveals the photo controls.

6. **Tap the Action button.** The Action button is the button on the left side of the menu bar. (If you don't see the menu bar, tap the screen.) iPhone 3G displays a list of actions you can perform.

7. **Tap Assign To Contact.** This displays a list of all of your contacts.

8. **Tap the contact you want to associate with the photo.** The Move and Scale screen appears.

9. **Drag the image so that it's positioned on the screen the way you want.**

10. **Pinch or spread your fingers over the image to set the zoom level you want.**

11. **Tap Set Photo.** iPhone 3G assigns the photo to the contact and returns you to the photo.

To assign a photo using the Contacts list, follow these steps:

1. **On the Home screen, tap the Contacts icon to open the Contacts list.**

2. **Tap the contact that you want to add a photo to.** Your iPhone 3G displays the contact's Info screen.

3. **Tap Edit to put the contact into Edit mode.**

4. **Tap Add Photo.** iPhone 3G displays a list of photo options.

5. **Tap Take Photo.** This activates the camera on the iPhone 3G.

6. **Frame the person on the screen, then tap the green camera button to take the photo.** The Move and Scale screen appears.

7. **Drag the image so that it's positioned on the screen the way you want.**

8. **Pinch or spread your fingers over the image to set the zoom level you want.**

9. **Tap Set Photo.** iPhone 3G assigns the photo to the contact and returns you to the Info screen.

10. **Tap Done.** Your iPhone 3G exits Edit mode.

Sending a photo via e-mail

More often than you'd think, it comes in really handy to be able to send photos from your iPhone 3G to someone's e-mail. This is particularly true if it's a photo you've just taken with your iPhone 3G camera, because then you can share the photo pronto, without having to trudge back to your computer. You also have the option of e-mailing an existing photo in one of your iPhone 3G photo albums.

Caution

Having the technology to e-mail a photo at your fingertips is wonderful, but bear in mind that your recipient doesn't see the photo in its natural state. Instead, your iPhone 3G shrinks the photo to 640 pixels wide by 480 pixels tall, a pale shadow of its original 1600-x-1200-pixel glory.

Here are the steps to follow to take a photo with the iPhone 3G camera and then e-mail it:

1. **On the Home screen, tap Camera.** The Camera application appears.

2. **Line up your subject and tap the Camera button to take the picture.**

3. **Tap the Camera Roll button.** The Camera Roll photo album appears.

4. **Tap the photo you just took.** A preview of the photo appears.

5. **Tap the Action icon.** The Action button is the button on the left side of the menu bar. (If you don't see the menu bar, tap the screen.) iPhone 3G displays a list of actions you can perform.

6. **Tap Email Photo.** The New Message screen appears and iPhone 3G embeds the photo in the body of the message.

7. **Choose your message recipient and type a Subject line.**

8. **Tap Send.** iPhone 3G sends the message and returns you to the photo.

If you have an existing image in one of your iPhone 3G's photo albums that you'd prefer to e-mail, follow these steps:

1. **On the Home screen, tap Photos.** The Photo Albums screen appears.

2. **Tap the photo album that has the image you want to send.**

3. **Tap the photo you want to send.** Your iPhone 3G opens the photo.

4. **Tap the Action button.** The Action button is the button on the left side of the menu bar. (If you don't see the menu bar, tap the screen.) iPhone 3G displays a list of actions you can perform.

5. **Tap Email Photo.** In the New Message screen that is displayed, the photo appears in the body of the message.

6. **Choose your message recipient and type a Subject line.**

7. **Tap Send.** iPhone 3G sends the message and returns you to the photo.

Note

In order to send a photo via e-mail, you must have a default e-mail account set on your iPhone 3G. See Chapter 4 for information about setting up a default e-mail account.

Sending a photo to another mobile phone

Most cell phones can't handle e-mail messages with attachments, but they can accept (and send) images embedded in text messages using a technology known as MMS, which stands for Multimedia Messaging Service.

Your iPhone 3G is just the opposite: It has no problem sending photos as e-mail message attachments, but it doesn't do the MMS thing, so you can't use it to send (or receive) text messages with embedded images.

The upshot of all this is that you can't use your iPhone 3G to send a photo to another (non-iPhone 3G) cell phone: The iPhone 3G can't send a photo in a text message, and the other cell phone can't accept a photo attached to an e-mail message.

So you're out of luck, right? Actually, no. Most cell phone carriers have a way of sending an image to customers without it needing to be sent from an MMS-capable phone. To do this, follow the instructions for e-mailing a photo in the previous section. However, instead of addressing the message to the contact's e-mail, you use the contact's mobile phone number followed by a specific address. Here are the general addresses to use for some common carriers:

- **Alltel.** *number*@message.alltel.com
- **AT&T.** *number*@mms.att.net
- **Bell Mobility (Canada).** *number*@1x.ball.ca
- **Boost Mobile.** *number*@myboostmobile.com
- **Cingular.** *number*@mms.mycingular.com
- **Fido (Canada).** *number*@fido.ca
- **Rogers Wireless (Canada).** *number*@pcs.rogers.com
- **Sprint/Nextel.** *number*@messaging.sprintpcs.com
- **T-Mobile.** *number*@tmomail.net
- **US Cellular.** *number*@mms.uscc.net
- **Verizon.** *number*@vzwpix.com
- **Virgin Mobile.** *number*@vmobl.com

In each case, replace *number* with your friend's cell number. For example, if your friend's number is 317-555-1234 and his carrier is AT&T, then enter 3175551234@mms.att.net into the address field of the e-mail. This sends the image to the person's carrier who then converts it to an MMS message and sends it to the phone that corresponds with the number at the beginning of the address.

Genius

Create a custom MMS e-mail address for any contact you use to send photos. In the Contacts list, tap your contact, tap Edit, tap Add new Email, tap Other, and then tap Add Custom Label. Tap mms, and then tap Save. In the Email box, tap the person's MMS address, tap Save, and then tap Done. Now when you send a photo, tap the blue + icon in the To line, tap your contact, and then tap the MMS address.

Receiving a photo on your iPhone 3G

As we said earlier, the iPhone 3G doesn't have MMS capability. This is unfortunate because it does throw a wrench into fully enjoying the photo features that it offers. Not to worry though, we know a way around the problem of receiving a photo as well.

When a phone that isn't MMS capable is sent a photo, it receives a text with a Web site at which it can view the image. On your iPhone 3G, enter the Web site into Safari and you are able to see the image. Problem solved.

Keep in mind, however, that these sites won't come up as hyperlinked in the text program. You also can't copy and paste on your iPhone 3G. This means that you actually have to take out a pen and paper and write down the site so you can manually enter it into Safari. I think that's just Apple's way of keeping us grounded and at one with the physical world.

Sending a photo to your Flickr account

If you have a Flickr account, you can send photos from your iPhone 3G by e-mail. Flickr gives you an e-mail address just for doing this. When you want to upload a photo to Flickr, all you do is attach it to an e-mail as described earlier in the chapter, and then enter the address given you by Flickr into the address field. You can find out the address to use this by going to the following page:

www.flickr.com/account/uploadbyemail

Starting a photo slide show

You can set up your own slide show on your iPhone 3G including transition effects and even background music. That really impresses people.

You can get the standard slide show up and sliding by following these steps:

1. **On the Home screen, tap the Photo icon to open the Photo application.**
2. **Tap the album that you want to use in your slide show.** This opens the album to reveal its photos.
3. **Tap the photo that you want to start with.** Your iPhone 3G opens the photo on the screen.
4. **Tap the Play icon.** The Play icon is the right-pointing arrow in the middle of the menu bar. (If you don't see the menu bar, tap the screen.) iPhone 3G starts the slide show.

To pause the show, tap the screen, and then tap Play again to resume the festivities.

Creating a custom photo slide show

Okay, the basic slide show is pretty neat, with its nice dissolve transitions. However, your iPhone 3G also offers a few settings for creating custom slide shows. Here's how to bring them up on your iPhone 3G:

1. **On the Home screen, tap the Settings icon.** This opens the Settings screen.

2. **Tap the Photos icon.** Your iPhone 3G displays the Photos screen, as shown in figure 6.11.

You get four settings to configure your custom slide show:

● **Play Each Slide For.** You use this setting to set the amount of time that each photo appears on-screen. Tap Play Each Slide For, and then tap a time: 2 Seconds, 3 Seconds (this is the default), 5 Seconds, 10 Seconds, or 20 Seconds.

● **Transition.** You use this setting to spec-ify the type of transition that your iPhone 3G uses between each photo. Tap Transition and then tap the type of tran-sition you prefer: Cube, Dissolve (the default), Ripple (*very* fun), Wipe Across, or Wipe Down.

6.11 Use your iPhone 3G's Photos screen to create a custom slide show.

● **Repeat.** This setting determines whether the slide show repeats from the beginning after the last photo is displayed. To turn on this setting, tap the Repeat switch to the On position.

● **Shuffle.** You use this setting to display the album photos in random order. To turn on this setting, tap the Shuffle switch to the On position.

Playing a slide show with background music

Here's is a little bonus that the iPhone 3G throws your way. Yes, you can wow them back home by running a custom slide show, but you can positively make their jaws hit the floor when you add a music soundtrack to the show! They'll be cheering in the aisles.

Here's how you do it:

1. **Press the Home button to display the Home Screen.**

2. **Tap the iPod icon to open the iPod application.**

3. **Tap the playlist and then tap a song to get the playlist going.**

4. **Press the Home button to return to the Home screen.**

5. **Tap the Photos icon to open the Photo application.**

6. **Tap a photo album, and then tap Play to start the slide show.** Your iPhone 3G runs the slide show, and all the while your music plays in the background.

Deleting a photo

If you mess up a photo using the camera, you should delete it before people think you have shoddy camera skills (because we all know it was the phone's fault, right?). Similarly, if your iPhone 3G contains a synced photo you don't need any more, you can delete it to reduce clutter in the photo album that holds it. Happily, you don't have to worry about this being a permanent deletion, either. The syncing process only goes from your computer to your 3G when it comes to photos that come from your computer. So even if you remove a photo from the iPhone 3G, it remains safe on your computer.

To delete a photo, follow these steps:

1. **Tap Photos in the Home screen.**

2. **Tap the photo album that has the image you want to blow away.**

3. **Tap the doomed photo.** Your iPhone 3G opens the photo.

4. **Tap the image to reveal the controls.**

5. **Tap the Trash icon.** The Trash icon is on the right side of the menu bar. (If you don't see the menu bar, tap the screen.) iPhone 3G asks you to confirm the deletion.

6. **Tap Delete Photo.** iPhone 3G tosses the photo into the trash, wipes its hands, and displays the previous photo in the album.

Watching YouTube Videos

As if all those iPod video shenanigans weren't enough, your iPhone 3G comes with a YouTube application right on the Home screen, so you can watch whatever video everyone's talking about, or just browse around for interesting finds.

YouTube videos tend to be in Flash, a video format that the iPhone doesn't recognize. However, many of YouTube's videos have been converted to a format called H.264 which is a much higher quality video format and is playable on your iPhone 3G. The YouTube application plays only these H.264 videos.

To fire up the YouTube application, press the Home button to return to the Home screen, then tap the YouTube icon.

Finding a YouTube video

YouTube's collection of talking cats, stupid human tricks, and TV snippets is vast, to say the least. To help you apply at least a bit of order to the YouTube chaos, your iPhone 3G organizes the YouTube application in a similar way to the iPod feature. That is, you get four browse buttons in the menu bar — Featured, Most Viewed, Bookmarks, and Search — and a More button that, when tapped, tosses up three more browse buttons: Most Recent, Top Rated, and History, as shown in figure 6.12.

Here's a summary of what each browse button does for you:

- **Featured.** Tap this button to display a list of videos picked by the YouTube editors. The list shows each video's name, star rating, popularity, and length.

- **Most Viewed.** Tap this to see the videos with the most views. At the top you can tap Today, This Week, and All. This

6.12 Use the browse button in the YouTube application's menu bar to locate and manage YouTube's videos.

chooses the top-viewed videos of today, this week, or of all time. At the bottom of the list you can tap Load 25 More, which loads 25 more Most Viewed.

Note To get more detailed information about a video, tap the blue More Info icon. The screen that appears gives you a description of the video, tells you when it was added, and shows a list of related videos.

- **Bookmarks.** Tap this button to see a list of videos that you've bookmarked as being favorites.

- **Search.** Tap this to display a Search text box. Tap inside the box, type a search phrase, and then tap Search. YouTube sends back a list of videos that match your search term.

- **Most Recent.** Tap here to see the videos that have been posted on YouTube most recently.

- **Top Rated.** Tap this button to display the videos that have the highest user ratings.

- **History.** Give this a tap to see the videos that you've viewed.

Note   Don't want someone passing by to know that you're addicted to lonelygirl15? We can't blame you. Tap the History button, tap Clear, and when your iPhone 3G asks you to confirm, tap Clear History.

Saving a video as a favorite

Just like finding a great site on the Web, finding a gem in the mountain of cut glass that is YouTube is a rare and precious thing. Chances are you'll want to play that video again later, but you can't always rely on it being in your History list, or being able to find it using the Search feature. Fortunately, the YouTube application saves you such frustration by enabling you to save a video as a bookmark. You can then run the video any time you want by tapping the Bookmarks button.

Follow these steps to create a bookmark for a video:

1. **In the YouTube application, locate the video you want to save.**
2. **Tap the video to start the playback.**
3. **Tap the Bookmark icon.** The Bookmark icon is on the left side of the playback controls. (If you don't see the controls, tap the screen.) iPhone 3G creates a bookmark for the video.

Sending a link to a video via e-mail

If you come across yet another amazing guitar player video that you simply must share with a friend, the YouTube application makes it easy by enabling you to send that person an e-mail message that includes the video address as a link. Here's how it works:

1. **In the YouTube application, locate the video you want to save.**
2. **Tap the video to start the playback.**

160

3. **Tap the Email icon.** The Email icon is on the right side of the playback controls. (If you don't see the controls, tap the screen.) iPhone 3G creates a new message with Funny Video as the subject and the YouTube address in the body.

4. **Choose your message recipient.**

5. **Modify the Subject line and body text as you see fit.**

6. **Tap Send.** iPhone 3G sends the message and returns you to the video.

Customizing the YouTube menu bar

The YouTube application comes with seven browse buttons, but it can only squeeze four of them at a time on the menu bar. If you use, say, the History button much more often than the Featured button (as we do), don't stand for it! You can customize the menu bar to show those browse buttons you use most often. Here are the steps required:

1. **In the YouTube application, tap More in the menu bar.**

2. **Tap Edit.** Your iPhone 3G displays the Configure screen, which shows all seven browse buttons, as shown in figure 6.13.

3. **Drag a browse button that you want to add to the menu bar and drop it on whatever existing menu bar browse button you want it to replace.** For example, if you want to replace the Featured browse button with History, drag the History button and drop it on Featured. Your iPhone 3G replaces the old browse button with the new one.

4. **Repeat Step 4 to add any of your other preferred browse buttons to the menu bar.**

5. **Tap Done to save the new menu bar configuration.**

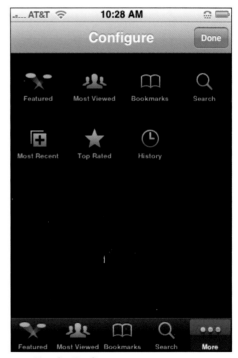

6.13 Use the Configure screen to create a custom YouTube menu bar.

161

Can I Use My iPhone 3G to Manage Contacts and Appointments?

The iPhone 3G has never been about just the technology. Yes, it looks stylish, has enough bells and whistles to cause deafness, and it just works, but iPhone 3G users don't know or care about things like antennas and flash drives and memory chips and whatever else Apple somehow managed to cram into that tiny case. These things don't matter because the iPhone 3G has always been about helping you get things done and helping you make your life better, more creative, and more efficient. And as you see in this chapter, your iPhone 3G can also go a long way toward making your life — particularly your contacts and your calendar — more organized.

Managing Your Contacts

One of the paradoxes of modern life is that as your contact information becomes more important, you store less and less of that information in the easiest database system of them all — your memory. That is, instead of memorizing phone numbers like you used to, you now store your contact info electronically. When you think about it, this isn't exactly surprising because it's not just a land-line number that you have to remember for each person; it might also be a cell number, an instant messaging handle, an e-mail address, a Web site address, a physical address, and more. That's a lot to remember, so it makes sense to go the electronic route. And for iPhone 3G, "electronic" means the Contacts list, which seems basic enough, but it's actually loaded with useful features that can help you organize and get the most out of the contact management side of your life.

Creating a new contact

We showed you how to sync your computer contacts program (such as Address Book on the Mac or Window, or Outlook's Contacts folder) in Chapter 5. That's by far the easiest way to populate your iPhone 3G Contacts list with a crowd of people, but it might not include everyone in your posse. If someone's missing and you're not around your computer, you can add that person directly to your iPhone 3G Contacts.

Here are the steps to follow:

1. **In the Home screen, tap the Contacts icon.** Your iPhone 3G opens the All Contacts screen. If you are in the phone application, you can also click the Contacts icon.

2. **Tap the + button at the top right of the screen.** The New Contact screen appears.

3. **Tap the First Last box.** Your iPhone 3G displays the Edit Name screen, as shown in figure 7.1.

7.1 When you tap the First Last field, your iPhone 3G displays the Edit Name screen so you can type the person's name and company name.

4. **The cursor starts off in the First box, so type the person's first name.** If you're jotting down the contact data for a company or some other inanimate object, skip to Step 7.

5. **Tap the Last box and then type the person's surname.**

6. **If you want to note where the person works (or if you're adding a business to your Contacts list), tap the Company box and type the company name.**

7. **Tap Save.** Your iPhone 3G saves the names and returns you to the New Contact screen.

Editing an existing contact

Now that your new contact is off to a flying start, you can go ahead and fill in details such as phone numbers, addresses (e-mail, Web, and real world), and anything else you can think of (or have the patience to enter into your iPhone 3G; it can be a lot of tapping!). The next few sections take you through the steps for each type of data. When you're done, be sure to tap Save to preserve all your hard work.

Note

There are a couple of techniques that we don't get into here because they've already been covered elsewhere. See Chapter 2 to learn how to assign a ringtone to your contact; see Chapter 6 to get the scoop on sprucing up your contact with a photo.

However, the steps we show also apply to any contact that's already residing in your iPhone 3G. Here, then, are the steps required to open an existing contact for editing:

1. **In the Home screen, tap the Contacts icon to open the All Contacts screen.**

2. **Tap the contact you want to edit.**

3. **Tap Edit.** Your iPhone 3G displays the contact's data in the Info screen.

4. **Make your changes, as described in the next few sections.**

5. **Tap Done.** Your iPhone 3G saves your work and returns you to the All Contacts screen.

Assigning phone numbers to a contact

Your iPhone 3G is, of course, a phone, so it's only right and natural to use it to call your contact. Sure, but which number? Work? Home? Cell? Pager? Fortunately, there's no need to choose just one, because your iPhone 3G is happy to store all these numbers, plus a few more if need be.

Here are the steps to follow to add one or more phone numbers for a contact:

1. **Tap Add new Phone.** Your iPhone 3G opens the Edit Phone screen, as shown in figure 7.2.

2. **With the cursor in the Phone field, type the phone number with area code first.** Your iPhone 3G helpfully adds extra stuff like parentheses around the area code and the dash.

3. **Examine the label box to see if the default label is the one you want.** If it is, skip to Step 5; if it's not, tap the label box to open the Label screen.

4. **Tap the label that best applies to the phone number you're adding (your iPhone 3G automatically sends you back to the Edit Phone screen after you tap):**

   - mobile
   - home
   - work
   - main
   - home fax
   - work fax
   - pager
   - other

7.2 Use the Edit Phone screen to assign a phone number to your contact.

Genius Some numbers — such as those used by long-distance calling cards — require a pause in mid-dial to wait for the system to do something. To tell your iPhone 3G to pause for two seconds while dialing, tap the +*# key, and then tap the Pause key. The iPhone 3G inserts a comma (,) to indicate the pause location.

5. **Tap Save.** Your iPhone 3G saves the changes and you end up back at the New Contact (or Info) screen with the new phone number displayed.

6. **Repeat Steps 1 to 5 to add any other numbers you want to store for this contact.**

Assigning e-mail addresses to a contact

It makes sense that you might want to add a phone number or three for a contact, but would you ever need to enter multiple e-mail addresses? Well, sure you would! Most people have at least a couple of addresses — usually home and work addresses — and some Type A e-mailers have a dozen or more. Life is too short to enter *that* many e-mail addresses, but you need at least the important ones if you want to use your iPhone 3G's Mail application to send a note to your contacts.

Follow these steps to add one or more e-mail addresses for a contact:

1. **Tap Add new Email.** Your iPhone 3G opens the Edit Email screen, as shown in figure 7.3.

2. **With the cursor in the Email field, type the person's e-mail address.** Note the handy @ and .com keys in the on-screen keyboard; you need those.

3. **Check out the label box to see if the default label is the one you want.** If it is, skip to Step 5; if it's not, tap the label box to open the Label screen.

4. **Tap the label that best applies to the e-mail address you're inserting (your iPhone 3G automatically sends you back to the Edit Email screen after you tap):**

   • **home**

   • **work**

   • **other**

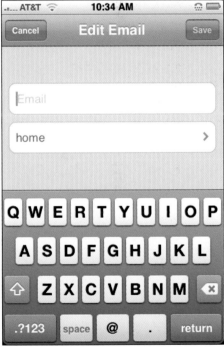

7.3 Use the Edit Email screen to add an e-mail address for your contact.

5. **Tap Save.** Your iPhone 3G saves the changes and sends you back to the New Contact (or Info) screen, where you see the new e-mail address.

6. **Repeat Steps 1 to 5 to add other e-mail addresses for this contact, as you see fit.**

Assigning Web addresses to a contact

Who on Earth *doesn't* have a Web site these days? It could be a humble home page, a blog, a home business site, or it could be someone's corporate Web site. Some busy Web beavers may even have all four! Whatever Web home a person has, it's a good idea to toss the address into his or her contact data because later on you can simply tap the address and your iPhone 3G (assuming it can see the Internet from here) immediately fires up Safari and takes you to the site. Does your pal have multiple Web sites? No sweat: Your iPhone 3G is happy to take you to them all.

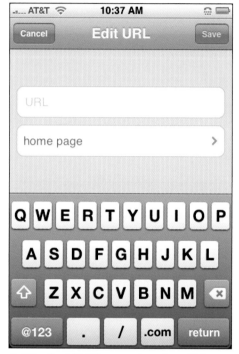

You can add one or more Web addresses for a contact by making your way through these steps:

1. **Tap Add new URL.** Your iPhone 3G opens the Edit URL screen, as shown in figure 7.4.

2. **With the cursor in the URL field, type in the person's Web address.** Note the . (period) and .com keys in the on-screen keyboard, which come in very handy.

7.4 Use the Edit URL screen to assign a Web address for your contact.

3. **Examine the label box to see if the default label is the one you want.** If it is, skip to Step 5; if it's not, tap the label box to open the Label screen.

Genius

To save some wear and tear on your tapping finger, don't bother adding the http:// stuff at the beginning of the address. Your iPhone 3G adds those characters automatically anytime you tap the address to visit the site. Same with the www. prefix. So if the full address is http://www.wordspy.com, you need only type wordspy.com.

4. **Tap the label that best applies to the Web address you're inserting (your iPhone 3G automatically sends you back to the Edit URL screen after you tap):**

    - **home page**

    - **home**

    - **work**

    - **other**

5. **Tap Save.** Your iPhone 3G saves the changes and sends you back to the New Contact (or Info) screen, where the new Web address is ready for action.

6. **Repeat Steps 1 to 5 to add other Web addresses for this contact.**

Assigning physical addresses to a contact

With all this talk about cell numbers, e-mail addresses, and Web addresses, it's easy to forget that people actually live and work *somewhere*. You may have plenty of contacts where the location of that somewhere doesn't much matter, but if you ever need to get from here to there, taking the time to insert a contact's physical address really pays off. Why? Because you need only tap the address and your iPhone 3G displays a Google map that shows you the precise location. From there you can get directions, see a satellite map of the area, and more. (We talk about all this great map stuff in Chapter 8.)

Tapping out a full address is a bit of work, but as the following steps show, it's not exactly root-canalishly painful:

1. **Tap Add new Address.** Your iPhone 3G opens the Edit Address screen, as shown in figure 7.5.

2. **Tap the first Street field and then type the person's street address.**

3. **If necessary, tap the second Street field, and then type even more of the person's street address.**

4. **Tap the City field, and then type the person's city.**

7.5 Use the Edit Address screen to tap out your contact's physical coordinates.

169

5. **Tap the State field, and then type the person's state.** Depending on what you later select for the country, this field might have a different name, such as Province.

6. **Tap the ZIP field, and then type the ZIP code.** Again, depending on what you later select for the country, this field might have a different name, such as Postal Code.

7. **Tap the Country to open the Country screen, and then tap the contact's country.**

8. **Examine the label box to see if the default label is the one you want.** If it is, skip to step 10; if it's not, tap the label box to open the Label screen.

9. **Tap the label that best applies to the Web address you're inserting (your iPhone 3G automatically sends you back to the Edit URL screen after you tap):**

   - **home**
   - **work**
   - **other**

10. **Tap Save.** Your iPhone 3G saves your work and returns you to the New Contact (or Info) screen, with the new address displayed.

11. **Repeat Steps 1 to 10 to add other addresses for this contact.**

Creating a custom label

When you fill out your contact data, your iPhone 3G insists that you apply a label to each tidbit: home, work, mobile, and so on. If none of the predefined labels fits, you can always just slap on the generic label: other. You *could* do that, but it seems so, well, *dull*. If you've got a phone number or address that you can't shoehorn into any of your iPhone 3G's prefab labels, get creative and make up a label. Here's how:

1. **Use either of the following techniques to get the show on the road:**

   - **If you want to apply the new custom label to an existing snippet of contact data, tap that data.**
   - **If you want to apply the new custom label to new data, tap the Add new *X* command (such as Add New Phone or Add new Email) that corresponds to the data you want to add.**

Note

To ensure that your iPhone 3G saves your custom label, you must actually edit or add some data and then save your work.

2. **Tap the label box.** The Label screen appears.

3. **Tap Edit.** Your iPhone 3G puts the Label screen into Edit mode.

4. **Tap Add Custom Label.** Scroll to the bottom of the screen to see this command. The Custom Label screen appears, as shown in figure 7.6.

5. **Type the custom label.**

6. **Tap Save.** Your iPhone 3G returns you to the screen for the field you were editing.

7. **Edit the rest of the field data.**

8. **Tap Save.** Your iPhone 3G saves the contact data as well as your custom label.

7.6 Use the Custom label screen to forge your very own custom label for your contacts.

Conveniently, you can apply your custom label to any type of contact data. For example, if you create a label named college, you can apply that label to a phone number, e-mail address, Web address, or physical address.

If a custom label wears out its welcome, follow these steps to delete it:

1. **Tap any data.**

2. **Tap the label box.** The Label screen appears.

3. **Tap Edit.** Your iPhone 3G puts the Label screen into Edit mode.

4. **Tap the red Delete icon to the left of the custom label you want to remove.** Your iPhone 3G displays a Delete button to the right of the field.

5. **Tap Delete.**

6. **Tap Done.** Your iPhone 3G deletes the custom label.

7. **Tap the Edit *X* button in the top-left corner.** *X* is the type of data you're editing (Phone, Web, and so on).

8. **Tap Cancel.** Your iPhone 3G returns you to the Info screen.

171

Adding extra fields to a contact

The New Contact screen (which appears when you add a contact) and the Info screen (which appears when you edit an existing contact) display just the fields you need for basic contact info. However, these screens lack quite a few common fields. For example, you might need to specify a contact's prefix (such as Dr. or Professor), suffix (such as Jr., Sr., or III), or job title.

Thankfully, your iPhone 3G is merely hiding these and other useful fields where you can't see them. There are 11 hidden fields that you can add to any contact:

- **Prefix**
- **Middle**
- **Suffix**
- **Phonetic First Name**
- **Phonetic Last Name**
- **Nickname**
- **Job Title**
- **Department**
- **Birthday**
- **Date**
- **Note**

The iPhone 3G is only too happy to let you add as many of these extra fields as you want. Here are the steps involved:

1. **Tap Add Field.** The Add Field screen appears, as shown in figure 7.7.
2. **Tap the field that you want to add.** Your iPhone 3G opens an Edit screen for that field.
3. **Enter the field data.**
4. **Tap Save.** Your iPhone 3G saves the new info and returns you to the Info (or New Contact) screen.

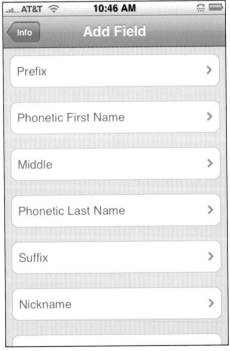

7.7 The Add Field screen shows the hidden fields that you can add to any contact.

Keeping track of birthdays and anniversaries

Do you have trouble remembering birthdays? If so, then we feel your pain because we, too, used to be pathetically bad at keeping birthdays straight. And no wonder: These days you not only have to keep track of birthdays for your family and friends, but increasingly often you have to remember birthdays for staff, colleagues, and clients, too. It's too much! Our secret is that we simply gave up and outsourced the job to our iPhone 3G's Contacts list, which has a hidden field that you can use to store birth dates.

To add the Birthday field to a contact, follow these steps:

1. **In the Contacts list, tap the contact you want to work with.**

2. **Tap Edit.** The Info screen appears.

3. **Tap Add Field.** Your iPhone 3G opens the Add Field screen.

4. **Tap Birthday.** The Edit Birthday screen and its nifty scroll wheels appear, as shown in figure 7.8.

5. **Scroll the left wheel to set the day of the month for the birth date.**

6. **Scroll the middle wheel to set the month for the birth date.**

7. **Scroll the right wheel to set the year of the birth date.**

8. **Tap Save.**

9. **Tap Done.** Your iPhone 3G saves the birthday info and displays it on the contact's Info screen.

7.8 Use the fun scroll wheels in the Edit Birthday screen to set the contact's birth date.

Everyone has a birthday, naturally, but lots of people have anniversaries, too. It could be a wedding date, a quit-smoking date, or the date that someone started working at the company. Whatever the occasion, you can add it to the contact info so that it's staring you in the face as a friendly reminder each time you open that contact.

Follow these steps to include an anniversary with a contact:

1. **In the Contacts list, tap the contact you want to edit.**

2. **Tap Edit.** Your iPhone 3G shows the Info screen.

3. **Tap Add Field.** The Add Field screen appears.

4. **Tap Date.** The Edit Date screen appears, as shown in figure 7.9.

5. **Scroll the left wheel to set the day of the month for the anniversary.**

6. **Scroll the middle wheel to set the month for the anniversary.**

7. **Scroll the right wheel to set the year of the anniversary.**

8. **The label box should already show the anniversary label, but if not, tap the label box, and then tap the anniversary.**

9. **Tap Save.**

10. **Tap Done.** The iPhone 3G saves the anniversary and displays it on the contact's Info screen.

7.9 Use the Edit Date screen to add an anniversary to a contact's info.

Note Although you can only add one birthday to a contact (not surprisingly), you're free to add multiple anniversaries. In the contact's Info screen, tap Edit, and then tap Add new Date.

Add notes to a contact

The standard contact fields all are designed to hold specific data: a name, an address, a date, and so on. Sometimes, however, you might need to enter more free-form data:

- The highlights of a recent client meeting.
- A list of things to do for the contact.

- How you met the contact, or why you added the person to your Contacts list.

- Contact data that doesn't have a proper field: spouse's or partner's name, kids' names, account numbers, gender, hobbies, and on and on.

Whatever it is, your iPhone 3G offers a Note field that you can add to a contact and then scribble away in as needed. To add the Note field to a contact, follow these steps:

1. **In the Contacts list, tap the contact you want to work with.**

2. **Tap Edit.** The Info screen appears.

3. **Tap Add Field.** Your iPhone 3G opens the Add Field screen.

4. **Tap Note.** The Edit Note screen appears, as shown in figure 7.10.

5. **Tap your notes into the large text box.**

6. **Tap Save.**

7. **Tap Done.** Your iPhone 3G saves the note text and displays it on the contact's Info screen.

7.10 Use the Edit Note screen to tap out notes related to the contact.

Creating a new contact from an electronic business card

Entering a person's contact data by hand is a tedious bit of business at the best of times, so it helps if you can find a faster way to do it. If you can cajole a contact into sending his or her contact data electronically, then you can add that data with just a couple of taps. What do we mean when we talk about sending contact data electronically? The world's contact management gurus long ago came up with a standard file format for contact data: the *vCard*. It's a kind of digital business card that exists as a separate file. People can pass this data along by attaching their (or someone else's) card to an e-mail message.

If you get a message with contact data, you see an icon for the VCF file, as shown in figure 7.11.

To get this data into your Contacts list, follow these steps:

1. **In the Home screen, tap Mail to open the Mail application.**

2. **Tap the message that contains the vCard attachment.**

3. **Tap the icon for the vCard file.** Your iPhone 3G opens the vCard.

4. **Tap Create New Contact.** If the person is already in your Contacts list, but the vCard contains new data, tap Add to Existing Contact, and then tap the contact.

Delete a contact field

People change, and so does their contact info. Most of the time these changes require you to edit an existing field, but sometimes people

7.11 If your iPhone 3G receives an e-mail message with an attached vCard, an icon for the file appears in the message body.

actually shed information. For example, they might get rid of their pager or fax machine, or they might shutter a Web site. Whatever the reason, you should delete that data from the contact to keep the Info screen tidy and easier to navigate.

To delete a contact field, follow these steps:

1. **In the Contacts list, tap the contact you want to work with.**

2. **Tap Edit.** The Info screen appears.

3. **Tap the red Delete icon to the left of the field you want to trash.** Your iPhone 3G displays a Delete button to the right of the field.

4. **Tap Delete.** Your iPhone 3G removes the field.

5. **Tap Done.** Your iPhone 3G returns you to the contact's Info screen.

Delete a contact

It feels good to add new contacts but, life being what it is, you don't get a lifetime guarantee with these things: friends fall out or fade away; colleagues decide to make a new start at another firm; clients take their business elsewhere; and some of your acquaintances simply wear out their welcome after a while. You move on, and so does your Contacts list, and the best way to do that is to delete the contact to help keep the list trim and tidy.

Follow these steps to delete a contact:

1. **In the Contacts list, tap the contact you want to get rid of.**

2. **Tap Edit.** The Info screen appears.

3. **Tap the Delete Contact button at the bottom of the screen.** Your iPhone 3G asks you to confirm the deletion.

4. **Tap Delete Contact.** Your iPhone 3G removes the contact and returns you to the All Contacts screen.

Tracking Your Appointments

When you meet someone and ask "How are you?," the most common reply these days is a short one: "Busy!" We're all as busy as can be these days, and that places-to-go-people-to-see feeling is everywhere. All the more reason to keep your affairs in order, and that includes your appointments. Your iPhone 3G comes with a Calendar application that you can use to create items, called *events*, which represent your appointments. Calendar acts as a kind of electronic personal assistant, leaving your brain free to concentrate on more important things.

Adding an appointment to your calendar

We showed you how to sync your computer's calendar application (such as iCal on the Mac, or Outlook's Calendar folder) in Chapter 5, and that's the easiest way to fill your iPhone 3G with your appointments. However, something always comes up when you're running around, so you need to know how to add an appointment directly to your iPhone 3G Calendar.

Here are the steps to follow to add a basic appointment:

1. **In the Home screen, tap the Calendar icon.** Your iPhone 3G opens the Calendar.

2. **Tap the date on which the appointment occurs.** If the appointment happens in a different month, use the arrow keys to the right and left of the month and year to navigate to the month you want.

3. **Tap the + button at the top right of the screen.** The Add Event screen appears, as shown in figure 7.12.

4. **Tap the Title/Location box.** You see the Title & Location screen.

5. **The cursor starts off in the Title box, so type a title for the appointment.** You can also tap the Location box and then type the location of the appointment.

6. **Tap Save.** Your iPhone 3G saves the data and returns you to the Add Event screen.

7. **Tap the Start/End box.** The Start & End screen appears, as shown in figure 7.13.

8. **Tap Starts, and then use the scroll wheels to set the date and time that your appointment begins.**

9. **Tap Ends, and then use the scroll wheels to set the date and time that your appointment finishes.**

Note If you're curious about the All-day setting, we talk about it later in this chapter.

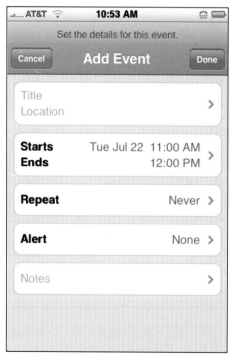

7.12 Use the Add Event screen to create your appointment.

7.13 Use the Start & End screen to set your appointment times.

10. **If you have multiple calendars, tap Calendar, and then tap the calendar in which you want this appointment to appear.**

11. **Tap Save.** Your iPhone 3G saves your info and returns you to the Add Event screen.

12. **Tap Done.**

When you add an appointment, Calendar displays a dot underneath the day as a visual reminder that you've got something going on that day. Tap the day and Calendar displays a list of all the events you've scheduled, as shown in figure 7.14.

If you want to see the duration of each appointment, tap the date and then tap Day. Calendar switches to Day view and shows your appointments as blocks, as you can see in figure 7.15.

7.14 Tap a date in Calendar to see all the events you've scheduled on that day.

7.15 Tap Day to switch to Day view and see your events as blocks of time.

Editing an existing appointment

Whether you've scheduled an appointment by hand or synced the appointment from your computer, the event details might change: a new time, a new location, and so on. Whatever the change, you need to edit the appointment to keep your schedule accurate.

Here are the steps to follow to edit an existing appointment:

1. **In the Home screen, tap the Calendar icon to open the Calendar.**

2. **Tap the date that contains the appointment you want to edit.**

3. **Tap the appointment.** You can do this either in Month view or in Day view. Calendar displays the event info.

4. **Tap Edit.** Your iPhone 3G displays the appointment data in the Edit screen.

5. **Make your changes to the appointment.**

6. **Tap Done.** Your iPhone 3G saves your work and returns you to the event details.

Setting up a repeating event

One of Calendar's truly great timesavers is the repeat feature, which enables you to set up a single event, and then get Calendar to automatically repeat the same event at a regular interval. For example, if you set up an event for a Friday, you can repeat the event every week, which means that Calendar automatically sets up the same event to occur on subsequent Fridays. You can continue the events indefinitely or end them on a specific date.

Follow these steps to configure an existing event to repeat:

1. **In Calendar, tap the date that contains the appointment you want to edit.**

2. **Tap the appointment.** Calendar opens the event info.

3. **Tap Edit.** Calendar displays the event data in the Edit screen.

4. **Tap Repeat.** The Repeat Event screen appears, as shown in figure 7.16.

5. **Tap the repeat interval you want to use.**

6. **Tap Save to return to the Edit screen.**

7.16 Use the Repeat Event screen to decide how often you want your event to recur.

7. **Tap End Repeat.** The End Repeat screen appears, as shown in figure 7.17.

8. **You have two choices here:**

   - **Have the event repeats stop on a particular day.** Use the scroll wheels to set the day, month, and year that you want the final event to occur, and then tap Save to return to the Edit screen.

   - **Have the event repeat indefinitely.** Tap Repeat Forever. Calendar returns you to the Edit screen.

9. **Tap Done.** Calendar saves the repeat data and returns you to the event details.

Converting an event to an all-day event

Some events don't really have specific times that you can pin down. These include birth-

7.17 Use the End Repeat screen to decide how long you want the even to repeat.

days, anniversaries, sales meetings, trade shows, conferences, and vacations. What all these types of events have in common is that they last all day: in the case of birthdays and anniversaries, literally so; in the case of trade shows and the like, "all day" refers to the entire work day.

Why is this important? Well, suppose you schedule a trade show as a regular appointment that lasts from 9AM to 5PM. When you examine that day in Calendar, you see a big fat block that covers the entire day. If you also want to schedule meetings that occur at the trade show, Calendar lets you do that, but it shows these new appointments "on top" of this existing trade show event. This makes the schedule hard to read, so you might miss an appointment.

To solve this problem, configure the trade show (or whatever) as an all-day event. Calendar clears it from the regular schedule and displays the event separately, near the top of the Day view. Here are the steps to follow:

1. **In Calendar, tap the date that contains the appointment you want to edit.**

2. **Tap the appointment.** Calendar opens the event info.

3. **Tap Edit.** Calendar switches to the Edit screen.

4. **Tap Starts/Ends.** The Start & End screen appears.

5. **Tap the All-day switch to the On position.**

6. **Tap Save to return to the Edit screen.**

7. **Tap Done.** Calendar saves the event and returns you to the event details.

Figure 7.18 shows Calendar in Day view with an all-day event added.

Adding an alert to an event

One of the truly useful secrets of stress-free productivity in the modern world is what we call the set-it-and-forget-it school of appoint-ments. That is, you set up an appointment elec-tronically, and then get the same technology

7.18 All-day events appear in the all-day section near the top of the Day view screen.

to remind you when the appointment occurs. That way, your mind doesn't have to waste energy fretting about missing the appointment because you know your technology has your back.

With your iPhone 3G, the technology of choice for doing this is Calendar and its alert feature. When you add an alert to an event, Calendar automatically displays a reminder of the event, which is a dialog box that pops up on the screen. Your iPhone 3G also vibrates and sounds a few beeps to get your attention. You also get to choose when the alert triggers (such as a specified number of min-utes, hours, or days before the event), and you can even set up a second alert just to be on the safe side.

Follow these steps to set an alert for an event:

1. **In Calendar, tap the date that contains the appointment you want to edit.**

2. **Tap the appointment.** Calendar opens the event info.

3. **Tap Edit.** Calendar displays the event data in the Edit screen.

4. **Tap Alert.** The Event Alert screen appears, as shown in figure 7.19.

5. **Tap the number of minutes, hours, or days before the event you want to see the alert.**

6. **Tap Save to return to the Edit screen.**

7. **To set up a backup alert, tap Second Alert, tap the number of minutes, hours, or days before the event you want to see the second alert, and then tap Save.**

8. **Tap Done.** Calendar saves your alert choices and returns you to the event details.

Figure 7.20 shows an example of an alert. Tap View Event to see the details, or tap OK to dismiss the alert.

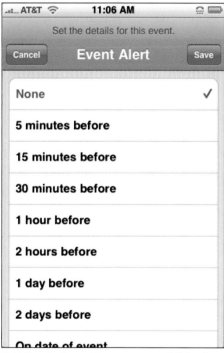

7.19 Use the Event Alert screen to tell Calendar when to remind you about your event.

7.20 Your iPhone 3G displays an alert similar to this when it's time to remind you of an upcoming event.

Caution If you flick the Ring/Silent switch on the side of the iPhone 3G to the Silent setting, remember that you won't hear the Calendar alert chirps. When the alert runs, your iPhone 3G still vibrates, and you still see the alert message on-screen.

Genius   You can disable the alert chirps if you find them annoying. On the Home screen, tap Settings, tap Sounds, and then tap the Calendar Alerts switch to the Off position.

Setting a birthday or anniversary reminder

If someone you know has a birthday coming up, you certainly don't want to forget! You can use your iPhone 3G's Contacts list to add a Birthday field for that person, and that works great if you actually look at the contact. If you don't, you're toast. The best way to remember is to get your iPhone 3G to do the remembering for you.

Follow these steps to set up a reminder about a birthday (or anniversary or some other important date):

1. **In Calendar, tap the date on which the birthday occurs.**
2. **Tap the + button at the top right of the screen.** The Add Event screen appears.
3. **Tap the Title/Location box, type a title for the event ("Karen's Birthday," for example), and then tap Save.** Calendar saves the data and returns you to the Add Event screen.
4. **Tap Starts/Ends, tap the All-day switch to the On position, use the scroll wheels to choose the birthday, and then tap Save.**
5. **Tap Repeat, tap Every Year, and then tap Save.**
6. **Tap Alert, tap the On date of event option, and then tap Save.**
7. **Tap Second Alert, tap the 2 days before option, and then tap Save.** This gives you a couple of days' notice, so you can go out and shop for a card and a present!
8. **Tap Done.** Calendar saves the event, and you have another load off your mind.

Displaying a list of your upcoming events

Calendar's Month view indicates dates that have scheduled events by displaying a teensy dot under the day number. So now you know you have *something* scheduled on those days, but the dots don't convey any more information than that. To see the scheduled events, you have to tap each day. *Way* too much work! A better way to is have Calendar do the work for you by displaying a list of what's scheduled over the next few days. Here's how:

1. **In the Home screen, tap the Calendar icon.** Your iPhone 3G opens the Calendar.

2. **Tap the List button.** Calendar displays a list of your upcoming events, as shown in figure 7.21. Tap an event to see its details.

Handling Microsoft Exchange meeting requests

If you've set up a Microsoft Exchange account in your iPhone 3G, there's a good chance you're using its push features, where the Exchange Server automatically sends incoming e-mail messages to your iPhone 3G, as well as new and changed contacts and calendar data. If someone back at headquarters adds your name to a scheduled meeting, Exchange generates an automatic meeting request, which is an e-mail message that tells you about the meeting and asks if you want to attend.

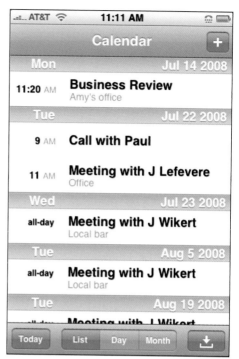

7.21 Tap Calendar's List button to see what's ahead on your schedule.

How will you know? Tap Calendar in the Home screen and then examine the bottom right of the screen. The menu bar icon that looks like an inbox tray will have a red dot with a number inside telling you how many meeting requests you've got waiting for you (see figure 7.22).

It's best to handle such requests as soon as you can, so here's what you do:

1. **Tap the inbox-like icon in the bottom right corner of the screen.** Calendar displays your pending meeting requests.

2. **Tap the meeting request you want to respond to.** Calendar displays the meeting details, as shown in figure 7.23.

185

Note

If you don't see the inbox tray icon, then you need to turn on syncing for your Exchange calendar. We show you how to do this in Chapter 4.

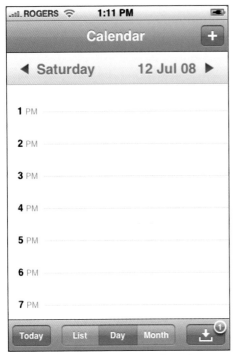

7.22 Calendar's meeting requests icon shows you how many Exchange meeting requests you've got.

7.23 The details screen for an Exchange meeting request.

3. **Tap your response:**

● **Accept.** Tap this button to confirm that you can attend the meeting.

● **Maybe.** Tap this button if you're not sure and will decide later.

● **Decline.** Tap this button to confirm that you can't attend the meeting.

Note

Meeting requests show up as events in your calendar, and you can recognize them thanks to their gray background. So another way to open the meeting details is to tap the meeting request in your calendar.

How Can I Use iPhone 3G to Help Organize My Life?

You'd be well within your rights to describe your iPhone 3G as a "Swiss Army phone," because it's positively bristling with tools: a phone, a Web browser, a media player, and a PDA (personal digital assistant). But your iPhone 3G also includes a broad range of features that help you organize and make sense of your life (some parts of it, anyway). We're talking here about even more tools: an alarm clock, a stopwatch, a map, a GPS device, a weather forecast, even a stock ticker! In this chapter, we explain these tools and show you how each one can help make small slices of your life easier and more productive.

Tracking Time with Clocks and Alarms

Ask anyone to come up with an Alternate Universe birthday wish list, and the item we're sure you'd see on almost everyone's list is "A 36-hour day." When you work longer hours, deal with a longer commute, and try to figure out how to incorporate aerobics and yoga classes, kids' parties and play dates, and whatever else is on the schedule, time becomes more precious than gold (or even gas!). What do you do when you sense that all the lanes in life's highway have become fast lanes, when it seems as though there are just not enough hours in the day to do everything you want or need to do, if you've abandoned *quality* time ages ago and would be happy to settle for a little *quantity* time?

You pull out your trusty iPhone 3G, of course! That may seem counterintuitive, but your iPhone 3G's Clock application actually has a few features that may not get you a 36-hour day, but they can help you manage the standard 24-hour day a bit more efficiently.

Seeing the time in another city

With your iPhone 3G in hand, tap the Home button to restore the Home screen, tap the Clock icon, and then tap World Clock in the menu bar. Your iPhone 3G displays the World Clock screen, which shows the current time in your location, as shown figure 8.1.

That's nice, but it's not much of a timesaver because your local time always appears in the iPhone 3G's status bar. Fortunately, the World Clock is no one-city wonder, and it's happy to show the current time in a whole bunch of cities. So a better use of the World Clock is to add clocks for cities you deal with regularly. For example, if you have associates in Arizona and its eccentric time zone (Mountain Standard Time) and no support for daylight–savings time means you can never figure out what time it is there, add a clock for Phoenix. Similarly, if you have homeboys in Hawaii, better add a clock for Honolulu so you don't end up calling someone in the middle of the night.

8.1 Your iPhone 3G's Clock application shows the current time in your neck of the woods.

Note We keep talking about adding *cities* to the Clock, and that's deliberate. The Clock application only knows about major centers, so if you have pals in Poughkeepsie, you need to add the New York clock.

To add a city to your World Clock, follow these steps:

1. **In the Clock application, tap the World Clock icon in the menu bar.** Clock displays the World Clock screen.

2. **Tap the + button in the top-right corner of the screen.** Clock displays a search box and a keyboard.

3. **Start typing the city you want to add.** As you type, cities that match the text appear on the screen, as shown in figure 8.2.

4. **When you see the city you're looking for, tap it.** Clock adds a clock for the city and returns you to the World Clock screen.

8.2 As you tap out some text in the search box, the Clock application shows the cities that match what you've typed so far.

Note As you type, you may see No Results Found instead of a list of cities. Double-check your typing to make sure you typed something that makes sense. If what you've typed looks okay, then we're afraid it means that your city isn't in Clocks database.

Figure 8.3 shows the World Clock screen with a few city clocks ticking away.

To keep your World Clock screen efficient and tidy, you can reorder the clocks and get rid of any you no longer need. In the World Clock screen, tap Edit, and then use the following techniques:

- **Change a clock's position in the list.** Slide the clock's drag bar up or down until the clock is in the position you prefer.

- **Remove a clock.** Tap the red Delete icon to the left of the clock, and then tap the Delete button that appears.

When your clock editing chores are complete, tap Done.

AT&T	12:24 AM	
Edit	**World Clock**	+
San Jose		**9:24** PM Yesterday
London		**5:24** AM Today
Paris		**6:24** AM Today
New York		**12:24** AM Today
World Clock	Alarm	Stopwatch	Timer

8.3 The World Clock screen with a few cities on the go.

Turning the iPhone 3G into an alarm clock

When you travel, you probably bring along an alarm clock so that you can rise and shine at an appropriate time if you have a breakfast meeting or a scuba lesson. Of course, the way the world works is that the more important the rendezvous, the more likely it is that you'll forget to pack your alarm clock. That's okay, though, because you already have an alarm clock with you: your iPhone 3G! The Clock application comes with a nifty Alarm feature that you can set to have your phone sound an alarm at whatever time you specify. It even has a snooze option, just in case you need an extra nine minutes of shut-eye before taking on the world.

Of course, no one is going to stop you if you want to use the Alarm feature for more than just help-ing you drag your weary body out of bed in the morning. Feel free to set an alarm whenever you need to be reminded about the time: when your lunch break is over, when a class is about to start, or when it's time to pick up the kids. Whatever's going on in your life, your iPhone 3G is there to help you show up on time.

Here are the steps to follow to set an alarm:

1. **In the Clock application, tap the Alarm icon in the menu bar.** Clock opens the Alarm screen.

2. **Tap the + button in the top-right cor-ner of the screen.** The Add Alarm screen appears, as shown in figure 8.4.

3. **If you want Clock to ring the alarm at the same time each week, tap Repeat to open the Repeat screen, then tap the repeat option you want: Every Monday, Every Tuesday, and so on.**

4. **Tap Sound to open the Sound screen, and then tap the sound that you want Clock to play when the alarm goes off.** If you'll be in a meeting or other public location where a sound is inappropriate, tap None to go soundless. Your iPhone 3G still vibrates when the alarm goes off, so you shouldn't miss it.

8.4 Use the Add Alarm screen to set your alarm.

Caution

Be sure to pick an alarm sound that will wake you up. Unless you're a *really* light sleeper, a mellow sound such as Harp or Timba just isn't going to do the job. You should also avoid the Crickets and Bark sounds, because you might mistake them for some local fauna! We also suggest using a sound that's different than your iPhone 3G's ringtone, otherwise you might mistake the alarm for an incoming call.

5. **If you want the option to delay the alarm for nine minutes, leave the Snooze switch in the On position.** If you don't need to snooze, set this switch to Off.

6. **To change the text that appears in the alarm message, tap Label, type the text, and then tap Back.**

7. **Flick the scroll wheels to set the hour and minute that you want the alarm to sound.**

8. **Tap Save.** Clock adds the alarm to the Alarm screen.

When the alarm time comes around, your iPhone 3G plays the sound, vibrates the phone, and displays the message, as shown in figure 8.5. Tap Snooze to set the alarm to sound again in nine minutes, or tap OK to dismiss it.

To keep your Alarm list in tip-top shape, you can edit any pending alarms and delete those you no longer need. In the Alarm screen, tap Edit, and then use the following techniques:

8.5 Your iPhone 3G displays a message similar to this when the alarm time comes around.

- **Modify an alarm.** Tap the alarm to open the Edit Alarm screen, make your changes, and then tap Save.

- **Remove an alarm.** Tap the red Delete icon to the left of the alarm, and then tap the Delete button that appears.

When you finish editing alarms, tap Done.

Note

The alarm tone is the one sound that your iPhone 3G does *not* suppress if you flick the Ring/Silent switch on the side of the phone to the Silent setting. The only way to silence an alarm is to set its Sound setting to None. In that case, your iPhone 3G still vibrates and you still see the alarm message on-screen.

Note There aren't many places where the iPhone 3G is a bit brain-dead, but the Lap feature of the stopwatch is one of them. Once you tap the Lap button, you can *only* track lap times. That is, there's no way to go back to see the full elapsed time.

Timing an event with the Stopwatch

You never know when you might stumble upon a track meet or a hotdog-eating contest, so it's good to know that you can always pull out your iPhone 3G and use its built-in stopwatch to time the action.

To use the stopwatch, follow these steps:

1. **In the Clock application, tap the Stopwatch icon in the menu bar.** Clock opens the Stopwatch screen, as shown in figure 8.6.

2. **Tap the green Start button to start timing.** The timer starts, the Start button turns to a red Stop button, and the Reset button turns to a Lap button.

3. **Tap the Lap key to record lap times.**

4. **Tap the Stop button to stop the timer.** The Stop button changes to the Start button. If you need to resume timing, tap Start.

8.6 Your iPhone 3G has a simple stopwatch built right in.

If another race is about to begin, tap the Reset button to clear the stopwatch display.

Caution When the stopwatch is running, the iPhone 3G will *not* go into Sleep mode. When you're done with the Stopwatch feature, press Home to exit Stopwatch and prevent your battery from draining.

195

Setting a countdown timer

The Clock application's Alarm feature is cool if you know what *time* you want the alarm to go off. However, there are lots of situations in life where you want to get a reminder a certain number of minutes or hours from *now*. Put an egg on to boil, and you might want a reminder in three or four minutes; decide to take a break from work, and you might want a reminder in 15 minutes. Your iPhone 3G can handle all this with ease thanks to its built-in Timer feature, which plays sounds, vibrates the phone, and displays a message after whatever interval you specify.

Timer has another great and useful trick: it can put your iPhone 3G into Sleep mode when the time interval runs out. That way, when you go to bed you can set your iPhone 3G to playing some nice, relaxing music, and you can rest easy knowing that the phone will shut itself off after a while. Sweet dreams!

Follow these steps to set the timer:

1. **In the Clock application, tap the Timer icon in the menu bar.** Clock opens the Timer screen, as shown in figure 8.7.

2. **Flick the scroll wheels to set the number of hours and minutes you want to use for the timer.**

3. **Tap When Timer Ends to open the When Timer Ends screen.**

4. **Tap the sound that you want Clock to play when the timer is finished.** If you want your iPhone 3G to go into Sleep mode, instead, tap Sleep iPhone.

5. **Tap Set to return to the Timer.**

6. **Tap Start.** Clock starts the timer and displays the countdown. Note that you can tap Cancel to bail out of the countdown.

When the timer hits 00:00, your iPhone 3G plays the sound, vibrates the phone, and displays the Timer Done message shown in figure 8.8. (If you chose Sleep iPhone rather than a sound, then your iPhone 3G goes into Sleep mode instead.) Tap OK.

Note

The timer sound still plays even if you flick the Ring/Silent switch on the side of your iPhone 3G to the Silent setting.

8.7 Tap Timer in the menu bar to set up a countdown timer on your iPhone 3G.

8.8 Your iPhone 3G displays this message when the timer counts down to 00:00.

Finding Your Way with Maps and GPS

When you're out in the real world trying to navigate your way between the proverbial points A and B, the questions often come thick and fast. "Where am I now?" "Which turn do I take?" "What's the traffic like on the highway?" "Can I even get there from here?" Fortunately, the answers to those and similar questions are now just a few finger taps away. That's because your iPhone 3G comes loaded not only with a way-cool Maps application brought to you by the good folks at Google, but it also has a global positioning system (GPS) receiver. Now your iPhone 3G knows exactly where it is (and so, by extension, you know where you are too), and it can help you get where you want to go.

Displaying your current location

When you arrive at an unfamiliar shopping mall and you need to get your bearings, your first instinct might be to seek out the nearest mall map and look for the inevitable "You Are Here" marker. This gives you a sense of your current location with respect to the rest of the mall, so locating The Gap shouldn't be all that hard.

When you arrive at an unfamiliar part of town or an unfamiliar city, have you ever wished your map could provide you with that same "You Are Here" reference point? If so, you're in luck because you've got exactly that waiting for you right in your iPhone 3G. Here's how it works:

1. **In the Home screen, tap the Maps icon.** The Maps application appears.

2. **Tap the Tracking button in the bottom- left corner.**

That's it! Your iPhone 3G examines GPS coordinates, Wi-Fi hot spots, and nearby cellular towers to plot your current position. When it completes the necessary processing and triangulating, your iPhone 3G displays a map of your current city, zooms in on your current area, and then adds a blue dot to the map to pinpoint your current location, as shown in figure 8.9. Amazingly, if you happen to be in a car, taxi, or other moving vehicle, the blue dot moves in real time!

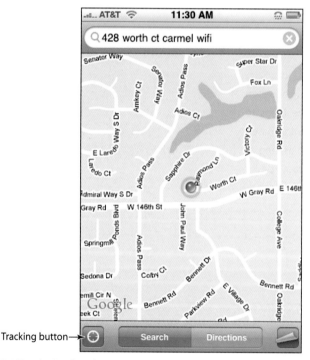

8.9 Tap the Tracking button to see your precise location as a blue dot on a map.

Genius

Knowing where you are is a good thing, but it's even better to know what's nearby. For example, suppose you're in a new city and you're dying for a cup of coffee. Tap Search in the menu bar, tap the Search box, type **coffee** (or perhaps *café* or *espresso*, depending on what you're looking for), and then tap Search. The Maps application drops a bunch of pins that correspond to nearby locations that match your search. Tap a pin to see the name, and tap the blue More Info icon to see the location's phone number, address, and Web site.

Displaying a map of a contact's location

In the old days (a few years ago!), if you had a contact located in an unfamiliar part of town or even in another city altogether, visiting that person required a phone call or e-mail asking for directions. You'd then write down the instructions, get written directions via e-mail, or perhaps even get a crudely drawn map faxed to you. Those days, fortunately, are long gone thanks to a myriad of online resources that can show you where a particular address is located and even give you driving directions to get there from here (wherever "here" may be).

Even better, your iPhone 3G takes it one step further and integrates with Google Maps to generate a map of a contact's location based on the person's contact address. So as long as you've tapped in (or synced) a contact's physical address, you can see where he or she is located on the map.

To display a map of a contact's location, follow these steps:

1. **In the Home screen, tap the Contacts icon to open the Contacts application.**

2. **Tap the contact you want to work with.** Your iPhone 3G displays the contact's data.

3. **Tap the address you want to map.** Your iPhone 3G switches to the Map applications and zeroes in on the contact's location. As you can see in figure 8.10, the contact's exact location is pinpointed — literally! — by a red pushpin.

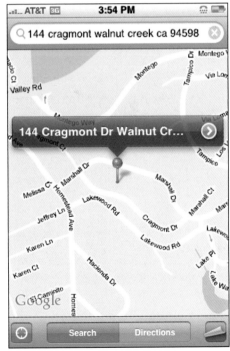

8.10 Tap a contact's physical address and your iPhone 3G uses the Maps application to display the location on a map.

199

Note   You can also display a map of a contact's location by using the Maps application itself. On the Home screen, tap Maps to open the Maps application. In the Search box (tap Search if you don't see this box; if the Search box contains text, tap X to delete it), tap the blue Bookmark icon on the right. Tap Contacts and then tap the contact you want to map. The Maps application maps the contact's address.

Saving a location as a bookmark for easier access

If you know the address of the location you want to map, you can add a pushpin for that location by opening the Maps application and running a search on the address. That is, you tap Search in the menu bar, tap the Search box, type the address, and then tap the Search button.

That's no big deal for one-time-only searches, but what about a location you refer to frequently? Typing that address over and over gets old in a hurry, we assure you. You can save time and tapping by telling the Maps application to save that location on its Bookmarks list, which means you can access the location usually with just a few taps.

Follow these steps to add a location to the Maps application's Bookmarks list:

1. **Search for the location you want to save.** The Maps application marks the location with a pushpin and displays the name or address of the location in a banner above the pushpin.

2. **Tap the blue More Info icon in the banner.** The Maps application displays the Info screen with details about the location:

   - If the location is in your Contacts list, you see the contact's data.

   - If the location is a business or institution, you see the address as well as other data such as the organization's phone number and Web address.

   - For all other locations, you see just the address.

3. **Tap Add to Bookmarks.** The Maps application displays the Add Bookmarks screen.

4. **Edit the name of the bookmark, if you want to, and then Tap Save.** The Maps application adds the location to the Bookmarks list.

To map a bookmarked location, follow these steps:

1. **Tap Search in the menu bar.**

2. **If the Search box has text, tap X to delete it.**

3. **Tap the blue Bookmark icon on the right side of the Search box.** The Maps application opens the Bookmarks screen.

4. **Tap Bookmarks in the menu bar.** The Maps application displays your list of bookmarked locations, as shown in figure 8.11.

5. **Tap the location you want to map.** The Maps application displays the appropriate map and adds a pushpin for the location.

8.11 You can access frequently used locations with just a few taps by saving them as bookmarks.

Specifying a location when you don't know the exact address

Sometimes you have only a vague notion of where you want to go. In a new city, for example, you might decide to head downtown and then see if there are any good coffee shops or restaurants. That's fine, but how do you get downtown from your hotel in the suburbs? Your iPhone 3G can give you directions, but it needs to know the endpoint of your journey, and that's precisely the information you don't have. Sounds like a conundrum, for sure, but there's a way to work around it. You can drop a pin on the map in the approximate area where you want to go. The Maps application can then give you directions to the dropped pin.

Here are the steps to follow to drop a pin on a map:

1. **In the Maps application, display a map of the city you want to work with:**

    - If you're in the city now, tap the Tracking icon in the lower-left corner of the screen.

    - If you're not in the city, tap Search, tap the Search box, type the name of the city (and per-haps also the name of the state or province), and then tap the Search button.

2. **Use finger flicks to pan the map to the approximate location you want to use as your destination.**

3. **Tap the Action button in the lower-right corner of the screen.** The Maps application dis-plays a list of actions.

4. **Tap Drop Pin.** The Maps application drops a purple pin in the middle of the current map.

5. **Drag the purple pin to the location you want.** The Maps application creates a temporary bookmark called Dropped Pin that you can use when you ask the iPhone 3G for directions.

Getting directions to a location

One possible navigation scenario with your iPhone 3G's Maps application is to specify a destina-tion (using a contact, an address search, a dropped pin, or a bookmark), then tap the Tracking but-ton. This gives you a map that shows both your destination and your current location. (Depending on how far away the destination is, you may need to zoom out — by pinching the screen or by tap-ping the screen with two fingers — to see both locations on the map.) You can then eyeball the streets to see how to get from here to there.

"Eyeball the streets"? Hah, how primitive! Your iPhone 3G's Maps application can bring you into the 21st century by not only showing you a route to the destination, but also by providing you with the distance and time it should take, *and* giving you street-by-street, turn-by-turn instructions. It's one of your iPhone 3G's sweetest features, and it works like so:

1. **Use the Maps application to add a pushpin for your journey's destination.** Use what-ever method works best for you: the Contacts list, an address search, a dropped pin, or a bookmark.

2. **Tap Directions in the menu bar.** The Maps application opens the Directions screen. As shown in figure 8.12, you should see Current Location in the Start box, and your destination address in the End box.

Swap button ——————

8.12 Use the Directions screen to specify the starting and ending points of your trip.

Note Instead of getting directions *to* the destination, you might need directions *from* the destination. No sweat. When you map the destination, tap the blue More Info icon and then tap Directions From Here. If you're already in the Directions screen, tap the Swap button to the left of the Start and End boxes. The Maps application swaps the locations.

3. **If you want to use a starting point other than your current location, tap the Start box and then type the address of the location you want to use.**

4. **Tap Route.** The Maps application figures out the best route and then displays it on the map in the Overview screen, which also shows the trip distance and approximate time.

Note Instead of seeing the directions one step at a time, you might prefer to see them all at once. Tap the Action icon in the lower-right corner of the screen, and then tap List.

5. **Tap Start.** The Maps application displays the directions for the first leg of the journey.

6. **Tap the Next (right arrow) key.** You see the directions for the next leg of the journey. Repeat to see the directions for each leg. You can also tap the Previous (left arrow) key to go back.

Genius

At the Address Info screen, if the address is in your Contacts list, the phone number is also displayed. Tap on the phone number to call the contact. You can let your friends know that you're on the way or you can make reservations at your favorite restaurant.

Getting live traffic information

Okay, it's pretty darn amazing that your iPhone 3G can tell you precisely where you are and precisely how to get somewhere else. However, in most cities it's the getting somewhere else part that's the problem. Why? One word: traffic. The Maps application might tell you the trip should take 10 minutes, but that could easily turn into a half hour or more if you run into a traffic jam.

That's life in the big city, right? Maybe not. If you're on a highway in a major U.S. city, the Maps application can most likely be able to supply you with — wait for it — real-time traffic conditions! This is really an amazing tool that can help you avoid traffic messes and find alternative routes to your destination.

To see the traffic data, tap the Action icon in the lower-right corner of the screen, and then tap Show Traffic. As you can see in figure 8.13, the Maps application uses four colors to illustrate the traffic flow:

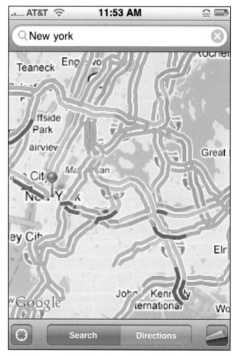

8.13 For most U.S. metropolitan highways, the color of the route tells you the current speed of the traffic.

- **Green.** Routes where the traffic is moving at 50 miles per hour or faster.
- **Yellow.** Routes where the traffic is moving between 25 and 50 mph.

- **Red.** Routes where the traffic is moving at 25 mph or slower.

- **Gray.** Routes that currently have no traffic data.

Now you don't have to worry about finding a news radio station and waiting for the traffic report. You can get real-time traffic information whenever you need it.

Viewing the Weather Forecast for Another City

If you're getting ready for a trip to another city, it's a good idea to check the current weather in that city. Why? Because if the weather in that city is radically different than your present weather, you need to decide what's best to wear while traveling. You should also check the forecast for the other city, because that helps you decide which clothes to pack.

It's normally a bit of a hassle to locate not only the current weather for a city, but also its weather forecast. Not so with your iPhone 3G, which comes with a Weather application that's incredibly quick and easy to use. You can add multiple cities, and for each city you see not only the current conditions, but also the six-day forecast.

To add a city to the Weather application:

1. **In the Home screen, tap the Weather icon to open the Weather application.**

2. **Tap the little i key in the lower-right corner of the screen.** This opens the Weather screen.

3. **Tap the + button in the top-left corner of the screen.** The All Locations screen appears.

4. **Type the city.** If you know a ZIP or postal code in the city, you can type that instead.

5. **Tap Search to see a list of matching cities.**

6. **Tap the city that you want.** Your iPhone 3G returns you to the Weather screen.

Note

If the forecast shows a blue background, it means that it's currently daytime in the city; if you see a purple background, instead, then it's currently nighttime in the city.

7. Tap either Celsius or Fahrenheit.

8. Tap Done. Your iPhone 3G displays the current conditions and the forecast for the city you added, as shown in figure 8.14.

You can add as many cities as you need, including, of course, your own! To scroll through the cities, just flick from one side of the screen to the other. If you want information about the city that's currently displayed in the Weather application, tap the Yahoo! icon in the bottom-left corner of the screen. This opens Safari with a page that has some information on the city.

You can also edit your cities by pressing the i key in the lower-right corner, and then using the following techniques:

- **Change a city's position in the list.** Slide the city's drag bar up or down until the city is in the position you prefer.

- **Remove a city.** Tap the red Delete icon to the left of the city, and then tap the Delete button that appears.

When your city editing is finished, tap Done.

8.14 Your iPhone 3G's Weather application shows the current conditions and the six-day forecast for whatever city you choose.

Writing and E-mailing a Note

What do you do if you're gallivanting around town and you suddenly think of a killer line for an upcoming speech, come up with a couple of good ideas for a project you're working on, or figure out the solution to some problem that's been rattling around in your head? In days of yore you'd probably pull out a piece of paper and a pen and scribble a note to yourself.

Paper? Pen? Scribble? Bah! It's an iPhone 3G world now, my friend, and you have no need for such ancient technology. Instead, you pull out your iPhone 3G and write a note to yourself the newfangled way:

1. **On the Home screen, tap the Notes icon.** Your iPhone 3G opens the Notes application. If you see the New Note screen right away (and you do if this is your first note), skip ahead to Step 3.

2. **Tap the + button at the top-right corner of the screen to open a new note.** The New Note screen appears, as shown in figure 8.15.

3. **Type your note.**

4. **Tap Done.** Your iPhone 3G puts away the keyboard and displays your note.

Having a note or two hanging out in your iPhone 3G is fine if all you're trying to do is remember some *bon mot* or good idea. However, what if it's a longer note and you actually want to use that text in some document, such as a word processing file on your computer? Your iPhone 3G doesn't offer any direct way to get the note to your computer (you can't sync your notes, for example), but it does offer an indirect route: e-mail it:

8.15 Use your iPhone 3G's Notes application to jot down a short missive to yourself.

1. **On the Home screen, tap the Notes icon to open the Notes application.**

2. **Tap the right and left arrow keys to navigate to the note you want.**

3. **Tap the envelope icon in the menu bar.** Your iPhone 3G creates a new e-mail message and inserts the note text into the body of the message and the subject.

4. **Tap the To box and type your e-mail address (or tap the blue + icon and select your contact).**

5. **Tap Send.** Your iPhone 3G e-mails the note.

To delete the note, display it in the Notes application and then tap the Trash icon in the menu bar. When Notes asks you to confirm, tap Delete Note to confirm.

Getting a Stock Quote

You can view the rise and fall of particular stocks throughout the day with your iPhone 3G's Stocks application. The stock data isn't real time — there's a 20-minute lag time — but that's standard for these kinds of free stock information services. You can track your own portfolio of stocks or stock indexes, and for each one you see the current price, how much the stock is up or down today, and a graph that shows the stock's price movements over the past year.

Here are the steps to follow to add a company or index to the Stocks application:

1. **In the Home screen, tap the Stocks icon.** Your iPhone 3G opens the Stocks application.

2. **Tap the i icon in the lower-right corner to open the Stocks screen.**

3. **Tap the + button.** The Add Stock screen opens and the keyboard appears.

4. **Type the company or index name or its abbreviation (if you know it).**

5. **Tap Search.** A list of matching companies or indexes appears.

6. **Tap the company or index that you want.** The stock is added to your list and you end up back on the Stocks screen.

7. **Tap either Numbers (if you want to see how much the stock has changed today) or % (if you want to see the percentage the stock has changed today).**

8. **Tap Done.** Your iPhone 3G displays the data for the company or index you added, as shown in figure 8.16.

8.16 Add a stock or index to see its data in the Stocks application.

You can edit your portfolio by pressing the i icon in the lower-right corner, and then using the following techniques:

- **Change a stock's position in the list.** Slide the stock's drag bar up or down until the stock is in the position you prefer.

- **Remove a stock.** Tap the red Delete icon to the left of the stock, and then tap the Delete button that appears.

When you're finished, tap Done.

How Do I Keep My Life In Sync with MobileMe?

When you go online, you take your life along with you, of course, so your online world becomes a natural extension of your real world. However, just because it's online doesn't mean the digital version of your life is any less busy, chaotic, or complex than the rest of your life. Apple's MobileMe service is designed to ease some of that chaos and complexity by automatically syncing your most important data — your e-mail, contacts, calendars, and bookmarks. Although the syncing itself may be automatic, setting up is not, unfortunately. This chapter shows you what to do.

From .Mac to MobileMe

These days, the primary source of online chaos and confusion is the ongoing proliferation of services and sites that demand your time and attention. What started with Web-based e-mail has grown to a Web site, a blog, a photo-sharing site, online bookmarks, and perhaps a few social networking sites, just to consume those last few precious moments of leisure time. You might be sitting in a chair, but you're getting run ragged anyway!

A great way to simplify your online life is to get a MobileMe account. For a Basic Membership fee ($99 per year currently), or a Family Pack membership, which consists of one main account plus four subaccounts ($149 per year currently), you get a one-stop Web shop that includes e-mail, an address book, a calendar, a Web Gallery for sharing photos, and a generous 20GB of online file storage (40GB with the Family Pack). The price is, admittedly, a bit steep, but it really is convenient to have so much of your online life in one place.

Note

If you don't want to commit any bucks before taking the MobileMe plunge, you can sign up for a 60-day trial that's free and offers most of the features of a regular account. Go to www.apple.com/mobileme/ and click Free Trial.

MobileMe is Apple's update to its .Mac online services, and if you're a .Mac member (or used it in the past) you'll find that MobileMe is very similar. You can still use your mac.com e-mail address, if you want, but Apple is issuing new addresses using the me.com domain, and is supplying existing members with an equivalent me.com address. (That is, if your old address was iPhoneFreak@mac.com, Apple automatically supplies you with the address iPhoneFreak@me.com.)

The Web applications that make up MobileMe — Mail, Contacts, Calendar, Gallery, and iDisk — are *much* nicer and much more functional than their .Mac predecessors. That's the "Me" side of MobileMe (because the Web applications are housed on Apple's me.com site), but that's not the big news with MobileMe. The real headline generator is the "Mobile" side of MobileMe. What's *mobile* is simply your data, particularly your e-mail accounts, contacts, calendars, and bookmarks. That data is stored on a bunch of me.com networked servers, which collectively Apple calls the *cloud*. When you log in to your MobileMe account at me.com, you use the Web applications to interact with that data.

That's pretty mundane stuff, right? What's revolutionary here is that you can let the cloud know about all the other devices in your life: your Mac, your home computer, your work PC, your notebook, and, of course, your iPhone 3G. If you log in to your MobileMe account and, say, add a new appointment, the cloud takes that appointment and immediately sends the data to all your devices. Fire up your Mac, open iCal, and the appointment's there; switch to your Windows PC, click

Outlook's Calendar folder, and the appointment's there; tap Calendar on your iPhone 3G's Home screen and, yup, the appointment's there, too.

This works if you change data on any of your devices. Move an e-mail message to another folder on your Mac, and the same message is moved to the same folder on the other devices and on your MobileMe account; modify a contact on your Windows PC, and the changes also propagate everywhere else. In each case, the new or changed data is sent to the cloud, which then sends the data to each device, usually in a matter of seconds. This is called *pushing* the data, and the new MobileMe applications are described as *push e-mail*, *push contacts*, and *push calendars*.

Note   If you've used e-mail, contacts, and calendars in a company that runs Microsoft Exchange Server, then you're no doubt used to push technology because Exchange has done that for a while through its ActiveSync feature (a feature that your iPhone 3G supports, by the way). MobileMe push is a step up, however, because you don't need a behemoth corporate server to make it happen. Apple calls MobileMe "Exchange for the rest of us."

With MobileMe, you never have to worry about entering the same information into all of your devices. With MobileMe, you won't miss an important meeting because you forgot to enter it into the calendar on your work computer. With MobileMe, you can never forget data when you're traveling because you have up-to-moment data with you at all times. MobileMe practically organizes your life for you; all you have to do is show up.

Understanding MobileMe Device Support

MobileMe promises to simplify your online life, but the first step to that simpler existence is to configure MobileMe on all the devices that you want to keep in sync. The next few sections show you how to configure MobileMe on various devices, but it's important to understand exactly which devices can do the MobileMe thing. Here's a summary:

- **iPhone.** MobileMe works with any iPhone that's running version 2.0 of the iPhone software.
- **Mac.** You must be running OS X 10.4.11 or later. To access the MobileMe Web applications, you need either Safari 3 or later, or Firefox 2 or later.
- **Windows XP.** You must be running Windows XP Service Pack 2 or later. To access the MobileMe Web applications, you need Internet Explorer 7 or later, Safari 3 or later, or Firefox 2 or later. For push e-mail you need either Outlook Express or Outlook 2003 or

later; for push contacts, you need either Windows Address Book or Outlook 2003 or later; for push calendar, you need Outlook 2003 or later.

- **Windows Vista.** Any Vista version works with MobileMe. To access the MobileMe Web applications, you need Internet Explorer 7 or later, Safari 3 or later, or Firefox 2 or later. For push e-mail you need either Windows Mail or Outlook 2003 or later; for push contacts, you need either Windows Contacts or Outlook 2003 or later; for push calendar, you need Outlook 2003 or later.

Configuring MobileMe on Your iPhone 3G

MobileMe is designed particularly with the iPhone 3G in mind, because it's when you're on the town or on the road when you need data pushed to you. To ensure your iPhone 3G works seamlessly with your MobileMe data, you need to add your MobileMe account and configure the iPhone 3G's MobileMe sync settings.

Setting up your MobileMe account on your iPhone 3G

Let's start with getting your MobileMe account set up on your iPhone 3G:

1. **On the Home screen, tap Settings.** Your iPhone 3G opens the Settings screen.

2. **Tap Mail, Contacts, Calendars.** The Mail, Contacts, Calendars screen appears.

3. **Tap Add Account.** The Add Account screen appears.

4. **Tap the MobileMe logo.** Your iPhone 3G displays the MobileMe screen, as shown in figure 9.1.

5. **Tap the Name text box and type your name.**

6. **Tap the Address text box and type your MobileMe e-mail address.**

9.1 Use the MobileMe screen to configure your MobileMe account on your iPhone 3G.

7. **Tap the Password text box and type your MobileMe password.** You can also tap the Description text box and type a short description of the account.

8. **Tap Save.** Your iPhone 3G verifies the account info and the MobileMe screen, as shown in figure 9.2.

9. **If you want to use push e-mail, leave the Mail switch set to On.**

10. **If you want to use push contacts, tap the Contacts switch to On and then tap Sync.**

11. **If you want to use push calendars, tap the Calendars switch to On and then tap Sync.**

12. **If you want to use push bookmarks, tap the Bookmarks switch to On and then tap Sync.**

13. **Tap Save.** Your iPhone 3G returns you to the Mail settings screen with your MobileMe account added to the Accounts list.

9.2 Use this MobileMe screen to activate push e-mail, contacts, calendars, and bookmarks.

Setting up MobileMe synchronization on your iPhone 3G

The "mobile" part of MobileMe means that no matter where you are, your e-mail messages, contacts, and calendars get pushed to your iPhone 3G and remain fully synced with all your other devices. Your iPhone 3G comes with this push feature turned on, but if you want to double-check this, or if you want to turn off push in order to concentrate on something else, you can configure the setting by following these steps:

1. **In the Home Screen, tap Settings.** The Settings screen appears.

2. **Tap Fetch New Data.** Your iPhone 3G displays the Fetch New Data screen, as shown in figure 9.3.

3. **If you want MobileMe data sent to you automatically, tap the Push switch to the On position.** Otherwise, tap Push to the Off position.

4. **If you turned push off, or if your iPhone 3G includes applications that don't support push, tap the frequency with which your iPhone 3G should fetch new data: Every 15 Minutes, Every 30 Minutes, Hourly, or Manually.**

Configuring MobileMe on Your Mac

If you want to keep your Mac in sync with MobileMe's push services, you need to add your MobileMe account to the Mail application and configure your Mac's MobileMe synchronization feature.

The first thing you need to do, if you haven't done so already, if download and install the Mac OS X Update for MobileMe. Here are the steps to follow:

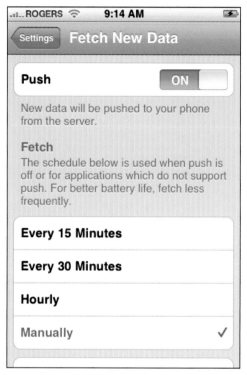

9.3 You use the Internet E-mail Settings to fill in most of the data you need for your MobileMe account.

1. **Pull down the Apple menu and choose Software Update.** The Software Update window appears and looks for updates.

2. **If you see Mac OS X Update for MobileMe, make sure its Install check box is activated, and then click Install 1 Item.** Your Mac asks for your password.

3. **Type your password, and then click OK.** Software Update installs the MobileMe update.

Setting up your MobileMe account on your Mac

Here are the steps to follow to get your MobileMe account into the Mail application:

1. **In the Dock, click the Mail icon.** The Mail application appears.

2. **Choose Mail ⇨ Preferences to open the Mail preferences.**

3. **Click the Accounts tab.**

4. **Click +.** Mail displays the Add Account dialog box.

5. **Type your name in the Full Name text box.**

6. **Type your MobileMe e-mail address in the Email Address text box.**

7. **Type your MobileMe password in the Password text box.**

8. **Leave the Automatically set up account check box selected.**

9. **Click Create.** Mail verifies the account info and then returns you to the Accounts tab with the MobileMe account added to Accounts list.

Setting up MobileMe synchronization on your Mac

Macs were made to sync with .Mac, so syncing with MobileMe should be a no-brainer. To ensure that's the case, you need to configure your Mac to make sure MobileMe sync is activated and that your e-mail accounts, contacts, and calendars are part of the sync process. Follow these steps to set your preferences:

1. **Click the System Preferences icon in the Dock.** Your Mac opens the System Preferences window.

2. **In the Internet & Network section, click the MobileMe icon.** The MobileMe preferences appear.

3. **Click the Sync tab.**

4. **Select the Synchronize with MobileMe check box.** Your Mac enables the check boxes beside the various items you can sync, as shown in figure 9.4.

5. **In the Synchronize with MobileMe list, choose Automatically.**

6. **Select the check box beside each data item you want to sync with your MobileMe account, particularly the following push-related items:**

   - **Bookmarks**

   - **Calendars**

   - **Contacts**

   - **Mail Accounts**

7. **Click the Close button.** Your Mac is now ready for MobileMe syncing.

217

9.4 Select the Synchronize with MobileMe check box and then select the items you want to sync.

Configuring Your MobileMe Account on Your Windows PC

MobileMe is happy to push data to your Windows PC. However, unlike with a Mac, your Windows XP or Vista machine wouldn't know MobileMe if it tripped over it. To get Windows hip to the MobileMe thing, you need to access your MobileMe account on me.com and configure it to work with your Windows PC.

Here are the steps to follow:

1. **On the Windows PC that you want to configure to work with MobileMe, select Start ⇨ Control Panel to open the Control Panel window.**

2. **Double-click the MobileMe Preferences icon.** If you don't see this icon, first open the Network and Internet category. If you still don't see it, you need to install the latest version of iTunes (at least 7.7). The MobileMe Preferences window appears.

3. **Use the Member Name text box to type your MobileMe member name.**

4. **Use the Password text box to type your MobileMe password.**

5. **Click Sign In.** Windows signs in to your account.

6. **Click the Sync tab.**

7. **Select the Sync with MobileMe check box and then choose Automatically in the Sync with MobileMe list, as shown in figure 9.5.**

8. **Select the Contacts check box, and then use the Contacts list to select the address book you want to sync.**

9. **Select the Calendars check box, and then use the Calendars list to select the calendar you want to sync.**

10. **Select the Bookmarks check box, and then use the Bookmarks list to select the Web browser you want to sync.**

11. **If you want to run a sync immediately, click Sync Now.**

12. **If you see the First Sync Alert dialog box, choose Merge Data, and then click Allow.**

13. **Click OK.**

9.5 Use the Configure Windows dialog box to set up your Windows PC to work with MobileMe.

Manually adding your MobileMe e-mail account to Outlook

In some scenarios, you might want to review your MobileMe e-mail messages with Outlook. In that case, you need to set up your MobileMe e-mail account by hand in Outlook. The next couple of sections show you how it's done in Outlook 2003 and Outlook 2007.

Setting up your MobileMe account in Outlook 2003

If you're a Microsoft Outlook 2003 user, follow these steps to configure your MobileMe account:

1. **Choose Tools ⇨ E-mail Accounts.** Outlook 2003 opens the E-mail Accounts Wizard.

2. **Select the Add a new e-mail account option, and then click Next.** The Server Type dialog box appears.

3. **Select the IMAP option, and then click Next.** The Internet E-mail Settings (IMAP) dialog box appears.

4. **Fill in the following data.** Figure 9.6 shows the dialog box filled in.

    - **Your Name.** Type your full name.

    - **E-mail Address.** Type your MobileMe address.

    - **Account Type.** Choose IMAP.

    - **Incoming mail server (IMAP).** Type **mail.me.com**.

    - **Outgoing mail server (SMTP).** Type **smtp.me.com**.

    - **User Name.** Type your MobileMe username (this should already be filled in for you).

    - **Password.** Type your MobileMe account password.

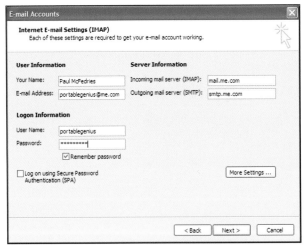

9.6 Use the Internet E-mail Settings (IMAP) dialog box to fill in most of the data you need for your MobileMe account.

5. **Click More Settings.** The Internet E-mail Settings dialog box appears.

6. **Click the Outgoing Server tab, and then select the My outgoing server (SMTP) requires authentication check box.** Be sure to leave the Use same settings as my incoming mail server option selected.

7. **Click the Advanced tab and change the following settings:**

    - **Incoming server (IMAP).** Change this port number to 143.

    - **Outgoing server (SMTP).** Change this port number to 587.

8. **Click OK to return to the wizard.**

9. **Click Next.** The final wizard page appears.

10. **Click Finish.**

Setting up your MobileMe account in Outlook 2007

If you're running Microsoft Outlook 2007, you need to follow these steps to configure your MobileMe account by hand:

1. **Choose Tools ⇨ Account Settings.** Outlook 2007 opens the Account Settings dialog box.

2. **In the E-mail tab, click New.** The Add New E-mail Account Wizard appears.

3. **Select the Microsoft Exchange, POP3, IMAP, or HTTP option, and then click Next.** The Auto Account Setup dialog box appears.

4. **Select the Manually configure server settings or additional server types check box, and then click Next.** The Choose E-mail Service dialog box appears.

5. **Select the Internet E-mail option, and then click Next.** The Internet E-mail Settings dialog box appears.

6. **Fill in the following data.** Figure 9.7 shows the dialog box filled in.

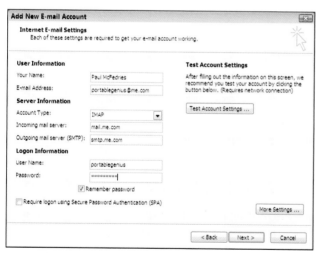

9.7 Use the Internet E-mail Settings dialog box to fill in most of the data you need for your MobileMe account.

- **Your Name.** Type your full name.

- **E-mail Address.** Type your MobileMe address.

- **Account Type.** Choose IMAP.

- **Incoming mail server.** Type **mail.me.com**.

- **Outgoing mail server (SMTP).** Type **smtp.me.com**.

- **User Name.** Type your MobileMe username (this should already be filled in for you).

- **Password.** Type your MobileMe account password.

7. **Click More Settings.** The Internet E-mail Settings dialog box appears.

8. **Click the Outgoing Server tab and then select the My outgoing server (SMTP) requires authentication check box.** Be sure to leave the Use same settings as my incoming mail server option selected.

9. **Click the Advanced tab and change the following settings:**

   - **Incoming server (IMAP).** Change this port number to 143.

   - **Outgoing server (SMTP).** Change this port number to 587.

10. **Click OK to return to the wizard.**

11. **Click Next.** The final wizard page appears.

12. **Click Finish to return to the Account Settings dialog box.**

13. **Click Close.**

Note

This is a good time to test the new account to make sure you've set it up properly. Click Test Account Settings. Outlook opens the Test Account Settings dialog box and attempts to log on to the incoming server and send a message through the outgoing server. If either test fails, check your settings.

Using Your iPhone 3G to Work with MobileMe Photos

Push e-mail, push contacts, and push calendars are the stars of the MobileMe show, and rightly so. However, the MobileMe interface at me.com also includes another Web application that shouldn't be left out of the limelight: the Gallery. You use this application to create online photo albums that you can share with other people, and even allow those folks to download your photos and upload their own.

You'll generally work with the Gallery either within the MobileMe interface on me.com, or by using compatible applications on your computer, such as iPhoto on your Mac. However, your iPhone 3G can also work with the Gallery, as you see in the next few sections.

Using your iPhone 3G to send photos to the MobileMe gallery

Your MobileMe account includes a Gallery application that you can use to create and share photo albums. You can upload photos to an album directly from the me.com site, or you can use iPhoto on your Mac to handle the upload chores.

However, what if you're cruising around town and use your iPhone 3G to snap a great photo of something? Do you really want to wait until you get back to your computer, sync the iPhone 3G, and *then* upload the photo? Of course not! Fortunately, you don't have to because you can send photos directly to your MobileMe Gallery right from your iPhone 3G.

Configuring an album to allow e-mail uploads

Before you can send those photos to your MobileMe Gallery, you have to configure the MobileMe album to allow e-mail photo uploads. Follow these steps:

1. **Use a Web browser to navigate to me.com and log in to your MobileMe account.**

2. **Click the Gallery icon to access the MobileMe Gallery.**

3. **Display the Album Settings dialog box:**

   - If you're creating a new photo album, click +.

   - If you want to use an existing album, click the album, and then click Settings.

4. **Select the Adding of photos via email or iPhone check box, as shown in figure 9.8.**

5. **If you want Gallery visitors to see the e-mail address used for sending photos to this album, select the Email address for uploading photos check box.**

6. **If you're creating a new album, type a name and configure the other settings as needed.**

7. **Click Publish.** If you're creating a new album, click Create, instead.

9.8 In the Album Settings dialog box, select the Adding of photos via email or iPhone check box.

Caution

If you configure your album so that anyone can see it, be careful about showing the e-mail address, otherwise your gallery could be invaded by irrelevant or even improper photos.

Note

If you selected the Email address for uploading photos check box, album visitors can see the upload address by clicking the Send to Album icon in your Web Gallery page.

Sending a photo to your own MobileMe Gallery

Now you're ready to send photos from your iPhone 3G directly to your MobileMe Gallery. Here's how it works:

1. **On the Home screen, tap Photos to open the Photos application.**

2. **Tap the photo album that contains the photo you want to upload.**

3. **Tap the photo.**

4. **Tap the Action icon in the lower-left corner.** Your iPhone 3G displays a list of actions.

5. **Tap Send to MobileMe.** Your iPhone 3G displays the MobileMe Albums screen.

6. **Tap the album you want to use for the photo.** Your iPhone 3G creates a new e-mail message. As you can see in figure 9.9, your iPhone 3G addresses the message to MobileMe and embeds the photo in the message body.

7. **Tap the Subject field and then edit the subject text.** This is the title that appears under the photo in the MobileMe Gallery.

8. **Tap Send.** Your iPhone 3G blasts the photo to your MobileMe Gallery.

9.9 An iPhone 3G photo ready to ship to your MobileMe Gallery.

Sending a photo to someone else's MobileMe Gallery

If you want to send a photo to another person's MobileMe Gallery, first check to make sure that e-mail uploads are allowed. Open the other person's Web gallery in any desktop browser (this won't work in your iPhone 3G's Safari browser), and then click Send to Album. If you see the Send to Album dialog box, note the e-mail address, and then click OK. (If nothing happens when you click Send to Album, it means the person doesn't want to share the address with the likes of us.)

Assuming you have the album upload address in your mitts, you can send a photo from your iPhone 3G to that person's MobileMe Gallery by following these steps:

1. **On the Home screen, tap Photos to open the Photos application.**

2. **Tap the photo album that contains the photo you want to upload.**

3. **Tap the photo.**

4. **Tap the Action icon in the lower-left corner.** Your iPhone 3G displays a list of actions.

5. **Tap Email Photo.** Your iPhone 3G displays the New Message screen.

6. **Tap the To field, and then type the other person's MobileMe Gallery upload e-mail address.**

7. **Tap the Subject field, and then edit the subject text.** This is the title that appears under the photo in the MobileMe Gallery.

8. **Tap Send.** Your iPhone 3G fires off the photo to the other person's Gallery.

Viewing your MobileMe Gallery in your iPhone 3G

Once you have an album or two lurking in your MobileMe Gallery, others can view your albums by using your special Gallery Web address, which takes the following form:

http://gallery.me.com/*username*

Here, *username* is your MobileMe username. For a specific album, the address looks like this:

http://gallery.me.com/*username*/#*nnnnnn*

Here, *nnnnnn* is a number that MobileMe assigns to you.

Naturally, because your Gallery is really just a fancy Web site, you can access it using your iPhone 3G's Safari browser, which also provides you with tools for navigating an album.

Here are the steps to follow to use your iPhone 3G to access and navigate a photo album in your MobileMe Gallery:

1. **On the Home screen, tap Safari.** Your iPhone 3G opens the Safari application.

2. **Tap the address bar to open it for editing.**

3. **Type your MobileMe Gallery address, and then tap Go.** The My Gallery page appears.

4. **Tap the album you want to view.** Safari displays thumbnail images for each photo.

5. **Tap the first photos you want to view.** Safari displays the photos as well as the controls for navigating the album, as shown in figure 9.10. If you don't see the controls, tap the photo.

9.10 Safari showing a photo from a MobileMe Gallery album.

6. **Tap the Next and Previous buttons to navigate the photos.** If you prefer a slide show, tap the Play button, instead.

Genius

If you check out your MobileMe Gallery frequently, save it as a bookmark for faster access. With the My Gallery page displayed, tap +, tap Add Bookmark, and then tap Save.

How Do I Enhance My iPhone 3G with App Store?

Your iPhone 3G is an impressive, eyebrow-raising device right out of the box. It does everything you want it to do, or so you think, but then you find out about some previously unknown feature and you wonder how you ever lived without it. It's hard to imagine that anyone would want to improve upon the iPhone 3G, or even that anyone *could* improve upon it. However, as you see in this chapter, the new App Store can make your iPhone 3G more convenient, more productive, and more, well, *anything!*

Accessing the App Store on Your Computer

You've seen that your iPhone 3G comes loaded with not only a basketful of terrific technology, but also a decent collection of truly amazing applications, all of which take advantage of the iPhone 3G's special features. But it won't escape your notice that the iPhone 3G's suite of applications is, well, incomplete. Where are the news and sports headlines? Why isn't there an easy way to post a short note to your blog, or a link to your de.licio.us account? And why on Earth isn't there a game in sight?

Fortunately, it's possible to fill in these and many other gaping holes in the iPhone 3G application structure. In the months before the release of the iPhone 3G, software developers from all over the world were busy cobbling together applications expressly designed to take advantage of iPhone 3G technologies such as the touchscreen, the accelerometer, GPS, and the tight links between iPhone 3G's Mail, Safari, Contacts, and Calendar applications. The result is a raft of high-quality applications in categories such as business, education, social networking, games, and many more.

Apple has taken all these applications and gathered them together in a new feature called the App Store. In the same way that you use the iTunes Store to browse and purchase songs and albums, you use the App Store to browse and purchase applications. (Although many of them are free for the downloading.) It's done using the familiar iTunes software on your Mac or Windows PC. (And, you can also connect to the App Store directly from your iPhone 3G, which is explained later.)

To access the App Store on your computer, follow these steps:

1. **Launch iTunes.**
2. **Click iTunes Store.** The iTunes Store interface appears.
3. **Click App Store.** iTunes loads the main App Store page, as shown in figure 10.1.

From here, use the links to browse the applications, or use the iTunes Store search box to look for something specific.

10.1 The main App Store page is the start of your search for iPhone 3G applications.

Downloading free applications

Early in the development process of the App Store, Apple made a pledge to the software developers: If you make your application free, then we won't charge you a cent to host it in the App Store. Getting to show off your digital handiwork in front of a few million people is the dream of any developer, but to get that access for nothing is almost too good to be true. *Almost.* The App Store does, indeed, boast a large collection of applications that are free for the downloading.

Here are the steps to follow to download and install a free application:

1. **In iTunes, click the App Store link.** Your computer opens the App Store for business.

2. **Use the App Store interface to locate the application you want to download.**

3. **Click the application.** The App Store displays a description of the application, along with its ratings, some screen shots, and some user reviews (see figure 10.2).

4. **Click Get App.** The App Store asks for your iTunes account password.

5. **Type your password, and then click Get.** iTunes downloads the application and stores it in the Library's Application category.

10.2 Click an application to see its details.

Purchasing applications

Even software developers have to make a living, so giving away applications might make good marketing sense, but it doesn't put Jolt Colas on the table in the short term. So, yes, many of the programs you see in the App Store will cost you a few dollars. That's okay if the application is decent, and hopefully you'll see a few reviews that let you know whether the application is worth shelling out the bucks.

Follow these steps to purchase and install a commercial App Store application:

1. **In iTunes, click the App Store link.** Your computer opens the App Store for business.

2. **Use the App Store interface to locate the application you want to download.**

3. **Click the application.** The App Store displays a description of the application. Pay particular attention to the application's rating and to the reviews that users have submitted.

4. **Click Buy App.** The App Store asks for your iTunes account password.

5. **Type your password and then click Buy.** iTunes downloads the application and stores it in the Library's Applications category.

Viewing and updating your applications

When you click Applications in the iTunes Library, you see a list of icons that represent all the applications that you've downloaded from the App Store, as shown in figure 10.3.

To check for updates to your apps, click Check for Updates. When the developer releases a new version of an application, App Store compares the new version with what you have. If you have an earlier version, it offers to update the application for you (usually without charge).

10.3 In the iTunes Library, click the Applications category to see your downloaded apps.

Moving Apps to Your iPhone 3G

After you download an application or two into iTunes, they won't do you much good just sitting there. To actually use the applications, you need to get them on your iPhone 3G. Here's how:

1. **Connect your iPhone 3G to your computer.** iTunes opens and accesses the iPhone 3G.
2. **In iTunes, click your iPhone 3G in the Devices list.**

3. **Click the Applications tab.**

4. **Select the Sync applications check box.**

5. **Select an option:**

   • **All applications.** Select this option to sync all your iTunes applications.

   • **Selected applications.** Select this option to sync only the applications you pick. In the application list, select the check box beside each application that you want to sync, as shown in figure 10.4.

6. **Click Apply.** iTunes syncs the iPhone 3G using your new applications settings.

10.4 You can sync selected applications with your iPhone 3G.

Accessing the App Store on Your iPhone 3G

Getting applications synced to your iPhone 3G from iTunes is great, but what if you're away from your desk and you hear about an amazing iPhone 3G game, or you realize that you forgot to download an important application using iTunes? This isn't even remotely a problem because your iPhone 3G can establish a wireless connection to the App Store anywhere you have Wi-Fi access or a cellular signal (ideally 3G for faster downloads). You can browse and search the applications, check for updates, and purchase any application you want (unless it's free, of course). The application downloads to your iPhone 3G and installs itself on the Home screen. You're good to go!

To access the App Store on your iPhone 3G, follow these steps:

1. **Tap the Home button to return to the Home screen.**

2. **Tap the App Store icon.**

As you can see in figure 10.5, your iPhone 3G organizes the App Store similar to the iTunes Store (as well as the iPod and YouTube applications). That is, you get five browse buttons in the menu bar — Featured, Categories, Top 25, Search, and Updates. You use these buttons to navigate the App Store.

Here's a summary of what each browse button does for you:

⬤ **Featured.** Tap this button to display a list of videos picked by the App Store editors. The list shows each application's name, icon, star rating, number of reviews, and price. Tap New to see the latest applications, and tap What's Hot to see the most popular items.

10.5 Use the browse buttons in the App Store's menu bar to locate and manage applications for your iPhone 3G.

Note   Tap an application to get more detailed information about it. The Info screen that appears gives you a description of the application, shows a screenshot, and may even offer some user reviews.

⬤ **Categories.** Tap this button to see a list of application categories, such as Games and Business. Tap a category to see a list of the applications available.

⬤ **Top 25.** Tap this button to see a list of the 25 most often downloaded applications.

- **Search.** Tap this button to display a Search text box. Tap inside the box, type a search phrase, and then tap Search. App Store sends back a list of applications that match your search term.

- **Updates.** Tap this button to install updated versions of your installed applications.

Downloading free applications

Amazingly, quite a few of the App Store applications cost precisely nothing. Nada. Zip. You might think that these freebies would be amateurish or too simple to be useful. While it's true that some of them are second-rate, a surprising number are full-fledged applications that are as polished and feature-rich as the commercial applications.

Follow these steps to download and install a free application:

1. **On the Home screen, tap App Store.** Your iPhone 3G opens the App Store.

2. **Locate the application you want to download, and then tap it.** The application's Info screen appears.

3. **Tap the Free icon.** The Free icon changes to the Install icon.

4. **Tap Install.** The App store asks for your iTunes account password.

5. **Type your password and then tap OK.** The App Store begins downloading the application. An icon for the application appears on the Home screen, and you see a progress bar that tracks the download and install process. (The icon title changes from Loading to Installing and finally to the name of the application itself.)

6. **When the installation is complete, tap the new icon on the Home screen to start using your new application.**

Note

If the application is quite big and you're surfing the Internet over a cellular connection — particularly an EDGE connection — your iPhone 3G may abort the installation and tell you that you need to connect to a Wi-Fi network to download the application.

Purchasing applications

Many of the iPhone 3G applications are extremely sophisticated, so it's not surprising that some of them will set you back a few bucks. To make sure you don't waste your money, read the description of the application, and be sure to read any reviews that other folks have submitted.

If a commercial application looks like something you want, follow these steps to purchase and install it:

1. **On the Home screen, tap App Store.** Your iPhone 3G connects to the App Store.

2. **Locate the application you want to purchase and then tap it.** The application's Info screen appears.

3. **Tap the price icon.** The price changes to a Buy Now icon.

4. **Tap the Buy Now icon.** The App Store asks for your iTunes account password.

5. **Tap the Password box, type your password, and then tap OK.** The App Store begins downloading the application. An icon for the application appears on the Home screen, and you see a progress bar that tracks the download and install process. (The icon title changes from Loading to Installing and finally to the name of the application itself.)

6. **When the installation is complete, tap the new icon on the Home screen to launch the application.**

Note App Store might not let you download a huge application if you're connected over a cellular signal. Instead of downloading the application, your iPhone 3G displays a message telling you to try again using a Wi-Fi connection.

Updating your applications

When you access the App Store with your iPhone 3G, take a look at the Updates browse button in the menu bar. If you see a red dot with a white number inside it superimposed over the Updates button, it means some of your installed applications have updated versions available. The number inside the dot tells you how many updates are waiting for you. It's a good idea to update your applications whenever a new version becomes available. The new version usually fixes bugs, but it might also supply more features, give better performance, or beef up the application's security.

Follow these steps to install an update:

1. **On the Home screen, tap App Store.** Your iPhone 3G connects to the App Store.

2. **Tap the Updates button.** Remember that you are only able to tap this button if you see the red dot with a number that indicates the available updates.

3. **Tap an update.** App Store displays a description of the update.

4. **Tap Update.** Your iPhone 3G downloads and installs the application update.

The good news about iPhone 3G problems — whether they're problems with iPhone 3G software or with the iPhone 3G itself — is that they're relatively rare. On the hardware side, although the iPhone 3G is a sophisticated device that's really a small computer (not just a fancy phone), it's far less complex than a full-blown computer, and so far less likely to go south on you. On the software side (and to a lesser extent on the accessories side), application developers (and accessory manufacturers) only have to build their products to work with a single device made by a single company. This really simplifies things, and the result is fewer problems. Not, however, *no* problems. Even the iPhone 3G sometimes behaves strangely or not at all. This chapter gives you some general troubleshooting techniques for iPhone 3G woes and also tackles a few specific problems.

General Techniques for Troubleshooting Your iPhone 3G

If your iPhone 3G is behaving oddly or erratically, it's possible that a specific component inside the phone is the cause, and in that case you don't have much choice but to ship your iPhone 3G back to Apple for repairs. Fortunately, however, most glitches are temporary and can often be fixed by using one or more of the following techniques:

- **Restart your iPhone 3G.** By far the most common solution to an iPhone 3G problem is to shut down and then restart the phone. By rebooting the iPhone 3G, you reload the entire system, which is often enough to solve many problems. You restart your iPhone 3G by pressing and holding the Sleep/Wake button for a few seconds, until you see the Slide to Power Off screen (at which point you can release the button). Drag the Slide to Power Off slider to the right to start the shutdown. When the screen goes completely black, your iPhone 3G is off. To restart, press and hold the Sleep/Wake button until you see the Apple logo, and then release the button.

- **Reboot your iPhone 3G's hardware.** When you restart your iPhone 3G by pressing and holding Sleep/Wake for a while, what you're really doing is rebooting the system software. If that still doesn't solve the problem, you might need to reboot the iPhone 3G's hardware, as well. To do that, press and hold down the Sleep/Wake button *and* the Home button. Keep them pressed until you see the Apple logo (it takes about eight seconds or so), which indicates a successful restart.

Genius The hardware reboot is also the way to go if your iPhone 3G is *really* stuck and holding down just the Sleep/Wake button doesn't do anything.

- **Recharge your iPhone 3G.** It's possible that your iPhone 3G just has a battery that's completely discharged. Connect your iPhone 3G to your computer or to the dock. If it powers up and you see the battery logo, then it's charging just fine and will be back on its feet in a while.

- **Shut down a stuck application.** If your iPhone 3G is frozen because an application has gone haywire, you can usually get it back in the saddle by forcing the application to quit. Press and hold the Home button for about six seconds. Your iPhone 3G shuts down the application and returns you to the Home screen.

- **Check for iPhone 3G software updates.** If Apple knows about the problem you're having, it will fix it and then make the patch available in a software update. We tell you how to update your iPhone 3G a bit later in this chapter.

- **Check for application updates.** It's possible that a bug in an application is causing your woes. On the Home screen, tap App Store, and then check the Updates button to see if any updates are available. If so, tap Updates, tap each application, and tap the Update button to make it so.

- **Erase and restore your content and settings.** This may seem like drastic advice, but it's possible to use iTunes to perform a complete backup of everything on your iPhone 3G. You can then reset the iPhone 3G to its original, pristine state, and then restore the backup. This rather lengthy process is explained later in the chapter.

- **Reset your settings.** Sometimes your iPhone 3G goes down for the count because its settings have become corrupted. In that case, you can restore the iPhone 3G by restoring its original settings. If iTunes doesn't recognize your iPhone 3G, then the backup-and-restore option is out. However, you can still reset the settings on the iPhone 3G itself. Tap Settings in the Home screen, tap General, tap Reset, and then tap Reset All Settings. When your iPhone 3G asks you to confirm, tap Reset All Settings.

Genius

If resetting the settings doesn't get the job done, it could be some recalcitrant bit of content that's causing the problem. In that case, tap Settings in the Home screen, tap General, tap Reset, and then tap Erase All Content and Settings. When your iPhone 3G asks you to confirm, tap Erase iPhone.

Troubleshooting connected devices

There are only a few ways that you can connect devices to your iPhone 3G: using the headset jack (which is much easier now that it's flush with the case, unlike the original iPhone), using the Dock connector, and using Bluetooth. So although the number of devices you can connect is relatively limited, that doesn't mean you might never have problems with those devices.

If you're having trouble with a device attached to your iPhone 3G, the good news is that a fair chunk of those problems have a relatively limited set of causes, so you may be able to get the device back on its feet by attempting a few tried-and-true remedies that work quite often for many devices. If it's not immediately obvious what the problem is, then your hardware troubleshooting routine should always start with these very basic techniques:

● **Check connections, power switches, and so on.** Some of the most common (and some of the most embarrassing) causes of hardware problems are the simple physical things: making sure that a device is turned on and checking that cable connections are secure. For example, if you can't access the Internet through your iPhone 3G's Wi-Fi connection, make sure your network's router is turned on, and make sure that the cable between your router and the ISP's modem is properly connected.

● **Replace the batteries.** Wireless devices such as headsets really chew through batteries, so if such a device is working intermittently or not at all, always try replacing the batteries to see if that solves the problem.

● **Turn the device off and then on again.** You *power cycle* a device by turning it off, waiting a few seconds for its innards to stop spinning, and then turning it back on again. You'd be amazed how often this simple procedure can get a device back up and running. For a device that doesn't have an on/off switch, try either unplugging the device from the power outlet, or try removing and replacing the batteries.

● **Reset the device's default settings.** If you can configure a device, then perhaps some new setting is causing the problem. If you recently made a change, try returning the setting back to its original value. If that doesn't do the trick, most configurable devices have some kind of Restore Default Settings option that enables you to quickly return the device to its factory settings.

● **Upgrade the device's firmware.** Some devices come with *firmware*, a small program that runs inside the device and controls its internal functions. For example, all routers have firmware. Check with the manufacturer to see if a new version exists. If it does, download the new version and then see the device's manual to learn how to upgrade the firmware.

Updating the iPhone 3G software

The iPhone 3G's software should update itself from time to time when you connect it to your computer, provided the computer has an Internet connection. This is another good reason to sync your iPhone 3G regularly. The problem is, you might hear about an important update that adds a feature you're really looking forward to or perhaps fixes a gaping security hole. What do you do if iTunes isn't scheduled to check for an update for a few days?

In that case, you take matters into your own hands and check for updates yourself:

1. **Connect your iPhone 3G to your computer.** iTunes opens and connects to your iPhone 3G.

2. **Click your iPhone 3G in the Devices list.**

3. **Click the Summary tab.**

4. **Click Check for Update.** iTunes connects to the Apple servers to see if any iPhone 3G updates are available. If an update exists, you see the iPhone Software Update dialog box, which offers a description of the update.

5. **Click Next.** iTunes displays the Software License Agreement.

6. **Click Agree.** iTunes downloads the software updates and then installs it.

Backing up and restoring the iPhone 3G's data and settings

Sometimes your iPhone 3G goes down for the count because its settings have become corrupted. In that case, you can restore the iPhone 3G by restoring its original settings. The best way to go about this is to use the Restore feature in iTunes, because that enables you to make a backup of your settings. However, it does mean that your iPhone 3G must be able to connect to your computer and be visible in iTunes.

If that's not the case, see our instructions for resetting in the next section. Otherwise, follow these steps to do a backup and restore on your iPhone 3G:

1. **Connect your iPhone 3G to your computer.**

2. **In iTunes, click your iPhone 3G in the Devices list.**

3. **Click Sync.** This ensures that iTunes has copies of all the data from your iPhone 3G.

4. **Click the Summary tab.**

5. **Click Restore.** iTunes asks if you want to back up your settings.

6. **Click Back Up.** iTunes asks you to confirm you want to restore.

7. **Click Restore.**

8. **If the iPhone Software Update dialog box appears, click Next, and then click Agree.** iTunes downloads the software, backs up your iPhone 3G, and then restores the original software and settings. When your iPhone 3G restarts, iTunes connects to it and displays the Set Up Your iPhone screen, as shown in figure 11.1.

9. **Select the Restore from the last backup of option.**

10. **If you happen to have more than one iPhone 3G backed up, use the list to choose yours.**

11.1 When your factory-fresh iPhone 3G restarts, use iTunes to restore your settings and data.

11. **Click Continue.** iTunes restores your backed-up data.

12. **Go through the tabs and check the sync settings to make sure they're set up the way you want.**

13. **Click Sync.** This ensures that your iPhone 3G has all of its data restored.

Note   The one thing that iTunes doesn't restore is your iPhone 3G's name. In iTunes' Devices list, double-click iPhone 3G, type the name you want to use, and then press Return or Enter.

Taking Care of the iPhone 3G Battery

Your iPhone 3G comes with a large lithium-ion battery, and Apple claims it gives you up to 5 hours of talk time (on a 3G network; 10 hours on an EDGE network); 5 hours of Internet use (on a 3G connection; 6 hours using Wi-Fi); 24 hours of audio playback; and 7 hours of video playback. Those are all impressive times, although count on getting less in the real world.

The biggest downside to the iPhone 3G battery is that it's not, in Apple parlance, a *user-installable* feature. If your battery dies, you have no choice but to return it to Apple to get it replaced. All the more reason to take care of your battery and try to maximize battery life.

Tracking battery use

Your iPhone 3G doesn't give a ton of battery data, but you can monitor both the total usage time (this includes all activities: calling, surfing, playing media, and so on) and standby time (time when your iPhone 3G was in Sleep mode). Here's how:

1. **On the Home screen, tap Settings.** The Settings screen appears.

2. **Tap General.** Your iPhone 3G displays the General options screen.

3. **Tap Usage.** Your iPhone 3G displays the Usage screen, as shown in figure 11.2.

4. **Examine the Usage value and the Standby value.** As your battery runs down, check the Usage screen periodically to get a sense of your iPhone 3G's battery use.

11.2 The iPhone 3G Usage screen

Tips for extending your battery life

Reducing battery consumption as much as possible on the iPhone 3G not only extends the time between charges, but it also extends the overall life of your battery. Here are a few suggestions:

- **Dim the screen.** The touch screen drains a lot of battery power, so dimming it reduces that power. On the Home screen, tap Settings, tap Brightness, and then drag the slider to the left to dim the screen.

- **Cycle the battery.** All lithium-based batteries slowly lose their charging capacity over time. If you can run your iPhone 3G on batteries for 4 hours today, later on you'll only be able to run it for 3 hours on a full charge. You can't stop this process, but you can delay it significantly by periodically *cycling* the iPhone 3G battery. Cycling — also called reconditioning or recalibrating — a battery means letting it completely discharge and then fully recharging it again. To maintain optimal performance, you should cycle your iPhone 3G's battery every one or two months.

Genius

Paradoxically, the *less* you use your iPhone 3G, the *more* often you should cycle its battery. If you often go several days or even a week or two without using your iPhone 3G (we can't imagine!), then you should cycle its battery at least once a month.

- **Slow the auto-check on your e-mail.** Having your e-mail poll the server for new messages eats up your battery. Don't set it to check every 15 minutes if possible. Ideally, set it to Manual check if you can. See Chapter 4 for information on how to do this.

- **Turn off push.** If you have a MobileMe account, consider turning off the push feature to save battery power. Tap Settings, tap Fetch New Data. In the Fetch New Data screen, tap the Push setting to Off and tap Manually in the Fetch section (see figure 11.3).

- **Minimize your tasks.** If you won't be able to charge your iPhone 3G for a while, avoid background chores such as playing music or secondary chores such as organizing your contacts. If your only goal is to read all your e-mail, stick to that until it's done because you don't know how much time you have.

- **Put your iPhone 3G into Sleep mode by hand, if necessary.** If you are interrupted — for example, the pizza delivery guy shows up on time — don't wait for your iPhone 3G to put itself to sleep because those few minutes use up precious battery time. Instead, put your iPhone 3G to sleep manually right away by pressing the Sleep/Wake button.

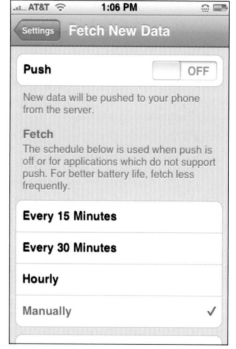

11.3 You can save battery power by turning off your iPhone 3G's push features.

- **Avoid temperature extremes.** Exposing your iPhone 3G to extremely hot or cold temperatures reduces the long-term effectiveness of the battery. Try to keep your iPhone 3G at a reasonable temperature.

Sending Your iPhone 3G in for Repairs

To get your iPhone 3G repaired, you could take your 3G to an Apple store or send it in. Visit www.apple.com/support and follow the prompts to find out how to send your iPhone 3G in for repairs. Remember that the memory comes back wiped, so be sure to sync with iTunes, if you can. Also, don't forget to remove your SIM card before you send it in.

- **Turn off Wi-Fi if you don't need it.** When Wi-Fi is on, it regularly checks for available wireless networks, which drains the battery. If you don't need to connect to a wireless network, turn off Wi-Fi to conserve energy. Tap Settings, tap General, tap Network, tap Wi-Fi, and then tap the Wi-Fi setting to Off.

- **Turn off GPS if you don't need it.** When GPS is on, the receiver exchanges data with the GPS system regularly, which uses up battery power. If you don't need the GPS feature for the time being, turn off the GPS antenna. Tap Settings, tap General, and then tap the Location Services setting to Off.

- **Turn off Bluetooth if you don't need it.** When Bluetooth is running, it constantly checks for nearby Bluetooth devices, and this drains the battery. If you aren't using any Bluetooth devices, turn off Bluetooth to save energy. Tap Settings, tap General, tap Bluetooth, and then tap the Bluetooth setting to Off.

Genius If you don't need all three of your iPhone 3G's antennas for a while, a faster way to turn them off is to switch your iPhone 3G to Airplane mode. Tap Settings, and then tap the Airplane Mode switch to the On position.

Solving Specific Problems

The generic troubleshooting and repair techniques that you've seen so far can solve all kinds of problems. However, there are always specific problems that require specific solutions. The rest of this chapter takes you through a few of the most common of these problems.

Your battery won't charge

If you find that your battery won't charge, here are some solutions:

- If the iPhone 3G is plugged into a computer to charge via the USB port, it may be that the computer has gone into standby. Waking the computer should solve the problem.

247

- The USB port might not be transferring enough power. For example, the USB ports on most keyboards don't offer much in the way of power. If you have your iPhone 3G plugged into a keyboard USB port, plug it into a USB port on the computer itself.

- Attach the USB cable to the USB power adapter, and then plug the adapter into an AC outlet.

- Double-check all connections to make sure everything is plugged in properly.

- Try an iPod cord if you have one.

If you can't seem to locate the problem after these steps, you may need to send your iPhone 3G in for service.

Caution

When you send in your iPhone 3G for any repair, the memory comes back wiped. Sync your iPhone 3G with your computer before you send it in to back up your contacts and other information. Don't forget to take out the SIM card as well.

You have trouble accessing a Wi-Fi network

Wireless networking adds a whole new set of potential snags to your troubleshooting chores because of problems such as interference and device ranges. Here's a list of a few troubleshooting items that you should check to solve any wireless connectivity problems you're having with your iPhone 3G:

- **Make sure the Wi-Fi antenna is on.** Tap Settings, tap Wi-Fi, and then tap the Wi-Fi switch to the On position.

- **Make sure the iPhone 3G isn't in Airplane mode.** Tap Settings and then tap the Airplane Mode switch to the Off position.

- **Check the connection.** The iPhone 3G has a tendency to disconnect from a nearby Wi-Fi network for no apparent reason. Tap Settings. If the Wi-Fi setting shows as Not Connected, tap Wi-Fi, then tap your network in the list.

- **Renew the lease.** When you connect to a Wi-Fi network, the access point gives your iPhone 3G a Dynamic Host Control Protocol (DHCP) lease that allows it to access the network. You can often solve connectivity problems by renewing that lease. Tap Settings, tap Wi-Fi, and then tap the blue More Info icon to the right of the connected Wi-Fi network. Tap the DHCP tab, and then tap the Renew Lease button, as shown in figure 11.4.

● **Reconnect to the network.** You can often solve Wi-Fi network woes by disconnecting from the network and then reconnecting. Tap Settings, tap Wi-Fi, and then tap the blue More Info icon to the right of the connected Wi-Fi network. Tap the Forget This Network button to disconnect, and then reconnect to the same network.

● **Reset your iPhone 3G's network settings.** This removes all stored network data and resets everything to the factory state, which might solve the problem. Tap Settings, tap General, tap Reset, and then tap Reset Network Settings. When your iPhone 3G asks you to confirm, tap Reset Network Settings.

● **Reboot and power cycle devices.** Reset your hardware by performing the following tasks, in order: restart your iPhone 3G, reboot your iPhone 3G's hardware, power cycle the wireless access point, power cycle the broadband modem.

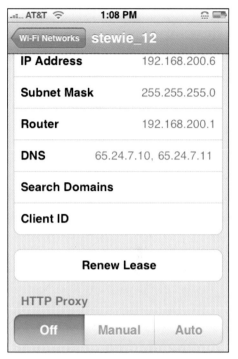

11.4 Open the connected Wi-Fi network's settings, and then tap Renew Lease to get a fresh lease on your Wi-Fi life.

● **Look for interference.** Devices such as baby monitors and cordless phones that use the 2.4 GHz radio frequency (RF) band can play havoc with wireless signals. Try either moving or turning off such devices if they're near your iPhone 3G or wireless access point.

● **Check your range.** If you're getting no signal or a weak signal, it could be that your iPhone 3G is too far away from the access point. You usually can't get much farther than about 115 feet away from an access point before the signal begins to degrade. Either move closer to the access point, or turn on the access point's range booster feature, if it has one. You could also install a wireless range extender.

You should keep your iPhone 3G and access point well away from a microwave oven; microwaves can jam wireless signals.

Caution

● **Update the access point firmware.** The access point *firmware* is the internal program that the access point uses to perform its various chores. Access point manufacturers frequently update their firmware to fix bugs, so you should see if an updated version of the firmware is available. See your device documentation to learn how this works.

● **Reset the router.** As a last resort, reset the router to its default factory settings (see the device documentation to learn how to do this). Note that if you do this you need to set up your network from scratch.

iTunes doesn't see your iPhone 3G

When you connect your iPhone 3G to your computer, iTunes should start and you should see the iPhone 3G in the Devices list. If iTunes doesn't start when you connect your iPhone 3G, or if iTunes is already running but the iPhone 3G doesn't appear in the Devices list, it means that iTunes doesn't recognize your iPhone 3G. Here are some possible fixes:

● **Check the connections.** Make sure the USB connector and the Dock connector are fully seated.

● **Try a different USB port.** The port you're using might not work, so try another one. If you're using a port on a USB hub, trying using one of the computer's built-in USB ports.

● **Restart your iPhone 3G.** Press and hold the Sleep/Wake button for a few seconds until the iPhone 3G shuts down, and then press and hold Sleep/Wake until you see the Apple logo.

● **Restart your computer.** This should reset the computer's USB ports, which might solve the problem.

● **Check your iTunes version.** You need at least iTunes version 7.7 to work with the iPhone 3G.

● **Check your operating system version.** On a Mac, your iPhone 3G requires OS X 10.4.11 or later; on a Windows PC, your iPhone 3G requires Windows Vista or Windows XP Service Pack 2 or later.

iTunes doesn't sync your iPhone 3G

If iTunes sees your iPhone 3G, but you can't get it to sync, you probably have to adjust some settings. See Chapter 5 for some troubleshooting ideas related to syncing.

You have trouble syncing music or videos

You may run into a problem syncing your music or videos to your iPhone 3G. The most likely culprit here is that your files are in a format that the iPhone 3G can't read. WMA, MPEG-1, MPEG-2, and other formats aren't readable to the iPhone 3G. First, convert them to a format that the iPhone 3G does understand using converter software. Then put them back on iTunes and try to sync again. This should solve the problem.

iPhone 3G-supported audio formats include Audible Formats 2, 3, and 4; Apple Lossless; AAC; AIFF; MP3; and WAV. iPhone 3G-supported video formats include H.264 and MPEG-4.

Your iPhone 3G doesn't recognize your SIM card

If your iPhone 3G doesn't detect your SIM card, try this:

1. **Eject the SIM card tray from the top of your phone using the tool that came with your iPhone 3G or a paper clip or pin.** Press the tool into the little hole on the tray and it should pop out.
2. **Make sure the SIM card is free of dirt and debris.**
3. **Reseat the SIM card in the tray and slide the tray back in.**

If this doesn't solve the problem, then your problem is a larger one and you need to contact Apple or your cellular provider.

Glossary

3G A third-generation cellular network that's faster than the old *EDGE* network, and enables you to make calls while also accessing the Internet.

802.11 See *Wi-Fi*.

accelerometer The component inside the iPhone 3G that senses the phone's orientation in space and adjusts the display accordingly (such as switching Safari from portrait view to landscape view).

access point A networking device that enables two or more devices to connect over a Wi-Fi network and to access a shared Internet connection.

ad hoc wireless network A wireless network that doesn't use an *access point*.

Airplane mode An operational mode that turns off the *transceivers* for the iPhone 3G's phone, Wi-Fi, and Bluetooth features, which puts the phone in compliance with federal aviation regulations.

authentication See *SMTP authentication*.

Bluetooth A wireless networking technology that enables you to exchange data between two devices using radio frequencies when the devices are within range of each other (usually within about 33 feet/10 meters).

bookmark An Internet site saved in Safari so that you can access the site quickly in future browsing sessions.

cloud The collection of me.com networked servers that store your MobileMe data and push any new data to your iPhone 3G, Mac, or Windows PC.

data roaming A cell phone feature that enables you to make calls and perform other activities such as checking for e-mail when you're outside of your provider's normal coverage area.

discoverable Describes a device that has its Bluetooth feature turned on so that other Bluetooth devices can connect to it.

double-tap To use a fingertip to quickly press and release the iPhone 3G screen twice.

EDGE (Enhanced Data rates for GSM [Global System for Mobile communication] Evolution) A cellular network that's older and slower than *3G*, although still supported by the iPhone 3G.

event An appointment or meeting that you've scheduled in your iPhone 3G's Calendar.

flick To quickly and briefly drag a finger across the iPhone 3G screen.

FM transmitter A device that sends the iPhone 3G's output to an FM station, which you then play through your car stereo.

GPS (Global Positioning System) A satellite-based navigation system that uses wireless signals from a GPS receiver — such as the one in the iPhone 3G — to accurately determine the receiver's current position.

group A collection of Address Book contacts. See also *smart group*.

headset A combination of headphones for listening and a microphone for talking.

Home screen The main screen on your iPhone 3G, which you access by pressing the Home button.

IMAP (Internet Message Access Protocol) A type of e-mail account where incoming messages, as well as copies of messages you send, remain on the server.

keychain A list of saved passwords on a Mac.

memory effect The process where a battery loses capacity over time if you repeatedly recharge it without first fully discharging it.

MMS (Multimedia Messaging Service) A technology that enables a cell phone to accept a text message with an embedded image.

multitouch A *touchscreen* technology that can detect and interpret two or more simultaneous touches, such as two-finger taps, spreads, and pinches.

pair To connect one Bluetooth device with another by entering a passkey.

pan To slide a photo or other image up, down, left, or right.

passcode A four-digit code that must be entered to unlock your iPhone 3G.

piconet An ad hoc wireless network created by two Bluetooth devices.

pinch To move two fingers closer together on the iPhone 3G screen. See also *spread*.

POP (Post Office Protocol) A type of e-mail account where incoming messages are only stored temporarily on the provider's mail server, and when you connect to the server, the messages are downloaded to iPhone 3G and removed from the server. See also *IMAP*.

power cycle To turn a device off, wait a few seconds for its inner components to stop spinning, and then turn it back on again.

preferences The options and settings, and other data that you've configured for your Mac via System Preferences.

push To send data immediately without being prompted.

ringtone A sound that plays when an incoming call is received.

RSS feed A special file that contains the most recent information added to a Web site.

silent mode An operational state where the iPhone 3G plays no sounds, except alerts set with the Clock application.

slide To drag a finger across the iPhone 3G screen.

smart group A collection of Address Book contacts where each member has one or more things in common, and where Address Book adds or deletes members automatically as you add, edit, and delete contacts.

smartphone A cell phone that can also perform other tasks such as accessing the Internet and managing contacts and appointments.

SMS (Short Message Service) A wireless messaging service that enables the exchange of short text messages between mobile devices.

SMTP (Simple Mail Transport Protocol) The set of protocols that determine how e-mail messages are addressed and sent.

SMTP authentication The requirement that you must log on to a provider's SMTP server to confirm that you're the person sending the mail.

SMTP server The server that an Internet service provider uses to process outgoing e-mail messages.

spread To move two fingers apart on the iPhone 3G screen. See also *pinch*.

SSID (Service Set Identifier) The name that identifies a network to *Wi-Fi* devices.

synchronization A process that ensures that data such as contacts, e-mail accounts, and events on your computer is the same as the data on your iPhone 3G.

tap To use a fingertip to quickly press and release the iPhone 3G screen.

touchscreen A screen that responds to touches such as finger taps and finger slides.

transceiver A device that transmits and receives wireless signals.

two-fingered tap To use two fingertips to quickly press and release the iPhone 3G screen.

vCard A file that contains a person's contact information.

.vcf The file extension used by a *vCard*.

wallpaper The background image you see when you unlock your iPhone 3G.

Web Clip A Home screen icon that serves as a link to a Web page that preserves the page's scroll position and zoom level.

Wi-Fi (Wireless Fidelity) A wireless networking standard that enables wireless devices to transmit data and communicate with other devices using radio frequency signals that are beamed from one device to another.

Index

The Genius is in.

978-0-470-29052-1

978-0-470-29050-7

978-0-470-38108-3

978-0-470-29169-6

978-0-470-29170-2

The essentials for every forward-thinking Apple user are now
available on the go. Designed for easy access to tools and shortcuts,
the *Portable Genius* series has all the information you need to maximize
your digital lifestyle. With a full-color interior and easy-to-navigate
content, the *Po_____ G_____* series _____ _____ _____ks as well
as savvy advi__ _____ _____ _____ _____uctivity.

Available wherever books are sold.

 WILEY

Now you know.